PROJECT MANAGEMENT
Principles and Practices

M. PETE SPINNER

Prentice Hall
Upper Saddle River, New Jersey   Columbus, Ohio

Library of Congress Cataloging-in-Publication Data

Spinner, M.
    Project management : principles and practices / M. Pete Spinner.
       p. cm.
    Includes index.
    ISBN 0-13-436437-6
    1. Industrial project management.   I. Title.
    HD69.P75S683   1997
    658.4'04—dc20

                                     96-11011
                                           CIP

Cover photo: © Bruce Peterson/H. Armstrong Roberts
Editor: Ed Francis
Production Editor: Linda Hillis Bayma
Design Coordinator: Jill E. Bonar
Text Designer: Susan E. Frankenberry
Cover Designer: Susan E. Frankenberry
Production Manager: Laura Messerly
Electronic Text Management: Marilyn Wilson Phelps, Matthew Williams, Karen L. Bretz,
    Tracey Ward
Marketing Manager: Danny Hoyt

This book was set in Leawood by Prentice Hall and was printed and bound by Quebecor
Printing/Book Press. The cover was printed by Phoenix Color Corp.

© 1997 by Prentice-Hall, Inc.
Simon & Schuster/A Viacom Company
Upper Saddle River, New Jersey 07458

Printed in the United States of America

10 9 8 7 6 5 4 3 2 1

ISBN: 0-13-436437-6

Prentice-Hall International (UK) Limited, *London*
Prentice-Hall of Australia Pty. Limited, *Sydney*
Prentice-Hall of Canada, Inc., *Toronto*
Prentice-Hall Hispanoamericana, S. A., *Mexico*
Prentice-Hall of India Private Limited, *New Delhi*
Prentice-Hall of Japan, Inc., *Tokyo*
Simon & Schuster Asia Pte. Ltd., *Singapore*
Editora Prentice-Hall do Brasil, Ltda., *Rio de Janeiro*

Preface

Books that have the term "project management" in their titles suggest that the contents encompass all of the nine major functions known as the project management body of knowledge: scope, time, costs, communications, human resources, quality, procurement, risk, and project integration. To cover these functions authoritatively and thoroughly in one book is a major undertaking. A single book would have to be quite voluminous to treat all these functions thoroughly. An alternative is to compromise by generalizing each function to produce a normal book size.

Project Management: Principles and Practices elects to use a third (and possibly a more desirable) option: it encompasses only the portions of project management that most persons need for (1) an introduction to the subject and (2) the ability to pursue their project work with the required basic principles. The book discusses only those functions of project management most requested by persons engaged in various forms of project management. *Project Management: Principles and Practices* therefore deals specifically with (1) planning and scheduling the timing and costs of a project, (2) maintaining control of timing and costs, (3) allocating labor and other personnel effectively over the course of the project, and (4) applying these principles to actual situations. The book also emphasizes that use of the computer is essential to effectively accomplishing the above.

This book starts with a brief, general discussion acquainting the reader with project management. It continues to explain how specific project management applications play an important role in producing successful projects. The term *project management* may be overused, as it may simply mean managing your own personal project in a disciplined manner. Similarly, a *project manager* or *project team* can be one person working on an individual project not requiring additional resources.

This book is intended to train the readers in basic project management principles for directing the course of a project. The hands-on approach presented in this book takes them through the necessary details for a good understanding of what to expect to complete a successful project. Often the readers themselves are not managing or directly participating in the project. Yet they still may have a role in its completion, and an understanding of the project is important. Users of this book will have an understanding, after following through the step-by-step stages, of how to plan and schedule projects. This systematic approach includes the application of project management software. The readers have a choice of using a software package that they have already acquired or using the basic software included with this book. (The software in the book includes simple step-by-step instructions for installation and for the applications.) The software is used in stages compatible with the subject being discussed. It enables readers to do computerized scheduling of timing, costs, and resources as they proceed from one chapter to the next. The computer also permits training in analyzing and resolving potential problems concerning project timing, cost scheduling, and allocation of labor/personnel.

The book is organized in two parts: Part I, consisting of the first seven chapters, deals with the necessary basic principles to plan and schedule a project, controlling the project by effectively applying the costs, labor/personnel, and other resources associated with it. Chapters 3 through Chapter 7 include computer applications for preparing timing and cost schedules and allocating labor and personnel. Exercises at the end of Chapter 3 through Chapter 7 contain situations requiring the use of the computer. Part II (Chapters 8–11) consists of case histories based on actual project experiences using the principles of project management explained and illustrated in Part I. These four case histories concern projects in engineering, construction, equipment procurement and installation, and product development. Businesses in these fields pioneered the use of planning and scheduling techniques beginning in the late 1950s.

In Chapter 8, the role of an architectural/engineering firm is examined as it prepares an engineering proposal for the building of a client's new fabrication facility. The firm will provide timing, costs, and resource information as it develops a plan and schedule, within budget, to meet the client's specific completion date for the new facility.

Chapter 9 concerns the planning and scheduling of a construction project. The building's owner requires the contractor and subcontractors to provide individual plans and schedules that will include timing, costs, and skilled labor in order to complete the project within specific time and cost restraints. This project involves use of basic project management principles that the contractor and subcontractors will need to understand.

In Chapter 10, personnel in a manufacturing facility need to expand and improve the paint finishing line of the company's fabricating plant. An industrial engineering and contracting firm has been retained to prepare an action plan to support the completion date and to formulate methods to control the project performance. This chapter illustrates how employing basic project management principles brings a project to successful completion.

Chapter 11 illustrates a product planning project using project management methods. The steps involved in bringing a new product from concept to market are discussed. The team members selected for the product planning project are responsible for preparing a plan and schedule for the product's development that satisfies a specific date for shipment. This chapter explains how each member of the team provides his or her own plan and schedule in conformance with the overall plan and schedule.

Project management principles have been increasingly applied to product development projects in recent years. Ford Motor Co. has enthusiastically and successfully introduced project management techniques in developing some of its new car models. The author, while at Ford, actively participated in employing project management for new car model introductions (as well as facility programs). These work assignments continued while working as a Ford consultant. The technique has become popular because of the notable results in reducing the cycle time for bringing new products to market. In turn, it has created a new area of learning for product designers and developers. Also discussed in Chapter 11 is how quality becomes an integral part of project management in the product planning process.

Much of the book's contents were used to train participants in the author's seminars and technology courses. The following educational and training institutions can benefit from use of this book:

- technical training schools that offer a basic project management or project development course within their computer-aided drafting and design curriculum
- training institutions offering programs that include construction engineering technology, industrial engineering technology, electrical engineering technology, and industrial design programs
- continuing education centers that offer basic and/or introductory project management courses
- colleges and universities offering civil, construction, and industrial engineering programs
- colleges and universities that do not offer hands-on project management courses, but need a reference book containing basic and/or introductory project management principles. The reference book would be suitable for engineering, marketing, and appropriate business courses

Acknowledgments

I thank People In Technology Limited, Epworth House, 25 City Road, London, England EC1Y1AA for permission to use the IN CONTROL! project management software to help explain the applications in *Project Management: Principles and Practices.* While People In Technology has more advanced project management "packages," I elected to use IN CONTROL! as it was the one most appropriate to initiate the reader with the computer applications in this book. People In Technology has several expansion and add-on programs for its Pertmaster and state-of-the-art Pertmaster Advanced and Pertmaster for Windows family that have made them one of the largest distributors of project management software in the United Kingdom. (Address any inquiries concerning Pertmaster for Windows to the London address.)

My appreciation also goes to those persons at Ford Motor Company whose contributions influenced specific portions of this book. To Winona Rhodes goes my appreciation for helping transform parts of the raw manuscript into a presentable package for submittal to the publisher. Finally I give a "thank you very much" to my wife, Margaret, for her patience throughout this endeavor. I will especially remember her being there during those numerous late-night hours, making copies, finding lost file material, and even emptying those frequently overfilled waste baskets.

M. Pete Spinner, PE, PMP

Contents

3

Project Schedule   33

4

Project Control   63

5

Project Costs: Schedule and Control   87

6

Personnel/Labor Planning   117

9

Planning and Scheduling a Building Project 195

10

A Facility Project: Measuring Project Performance 223

11
....

Product Development Project: Create Milestones   259

APPENDIX A:

Additional Instructions for IN CONTROL! Software   277

APPENDIX B:

Greenfield Facility Description   293

APPENDIX C:

Concept-to-Market (CTM) Project Management Seminar Outline   299

INDEX   303

BASIC PRINCIPLES OF PROJECT MANAGEMENT

Introduction

THIS CHAPTER WILL COVER

- Defining project management
- Importance of project management
- Where project management is used
- Benefits of project management

About 20 years ago, a project engineer quipped that the foremost "principles" of project management were "patience, understanding and wisdom." No matter how they were characterized, these personal attributes are still important in many areas of endeavor, not necessarily limited to project management. To complement these "principles," various tools and techniques that are relatively easy to learn are now available.

DEFINING PROJECT MANAGEMENT

What is project management, and what is a project? What is the background of project management—where should it be used? This chapter should help explain the benefits of applying project management to handling projects.

A **project** consists of a series of tasks (or activities) that have several distinguishing characteristics:

- The project has specific starting dates and ending dates.
- It has well-defined objectives.
- It achieves a specified product or result.
- It is a unique, nonrepetitive endeavor.
- Cost, time schedules, and resources (personnel/labor, equipment, material) are consumed.

Projects exist in nearly every field of endeavor, including designing and/or constructing a house, manufacturing facility, or commercial building; designing a new product; introducing a new product into the marketplace; installing new facilities for education, banking, health care, commercial, and industrial ventures; relocating a business or manufacturing process; relocating offices; and introducing a new business or software product.

Project management is defined as managing and directing time, material, personnel/labor, and costs to complete a project in an orderly, economical manner and to meet the established objectives of time, costs, and technical and/or service results.

A successful project using the project management approach will consist of three stages:

- **Planning**: understanding *what* work has to be done—including identifying the individual activities and necessary resources—to complete the project; then developing a plan of action in a logical order that can be displayed graphically as a project planning diagram (or network).
- **Scheduling**: validating *when* the work activities need to be done. This stage details the time allowed for the project and each activity—when they are to be started and completed.
- **Controlling**: monitoring (or tracking) progress of the project as it gets under way, analyzing performance, then resolving concerns. It also manages status reporting.

MANAGEMENT PRACTICES

Several significant management practices are instinctively used when applying project management methods. While numerous books have been written about them, our use requires just a modest understanding of their applications. A brief description of each of these principles as related to our use follows:

- **Network analysis** (also called graphic analysis) shows the plan of action through the use of a graphic diagram (used in project planning for preparing the planning diagram).
- **Management by objectives** is a technique that defines objectives and uses a disciplined method to measure performance against those objectives (used in project planning and project controlling).
- **Management by exception** is a technique that signals the specific problems requiring management attention. Studies indicate that management should be involved with no more than 20 percent of the total project at any one time (used extensively in project control).

Other management practices that are also very important functions of project management include:

- **Cost minimizing**: a procedure used to reduce the time required to complete a project with the least amount of additional cost.
- **Resource allocation**: assigning resources required to complete each project activity. Resources include personnel, labor, cost allocation (assigning costs required to complete each activity), equipment, and material.
- **Resource leveling**: a method for scheduling the project activities to keep day-to-day resource requirements even.

HISTORY OF PROJECT MANAGEMENT

While the procedures associated with project management first saw use in the United States in the 1950s, the crux of project management—the project planning diagram—traces its origins back to the 1850s, when students in logic and algebra used diagrams to show the flow of logical problems. The military-minded Prussians in that same century developed flow diagrams of tactical movements in the battlefield. They showed where their troops would be, where the enemy troops would be, and how the battle would progress.

In the late 1950s, E. I. duPont de Nemours and Co. and the U.S. Navy completed early work in project management combining a project planning diagram with a computerized schedule. Working on separate projects with separate teams, the projects they used to initiate their studies had little in common. Both had the same objectives, however: to improve the planning, scheduling, and coordination of their projects. They both recognized the need for improved planning and scheduling techniques to help control their use of personnel, labor, material and costs of resources.

DuPont's system was called the **critical path method (CPM)** with its objectives to improve efficiencies, timing, and costs of the company's engineering projects. This system helped reduce the timing of many projects by one-third with significant reduction in costs. Simultaneously, the Navy was devising a system to plan and coordinate the work of nearly 3,000 contractors and agencies engaged in producing the Polaris missile for launching nuclear warheads from submarines (a consequence of the Cold War with Russia). This system, known as the **Program Evaluation Review Technique (PERT)**, has been credited with helping advance the Polaris' development by two years.

With the success of these programs came a proliferation of project diagramming methods, including PEP (Project Evaluation Procedure), LESS (Least Cost Estimating and Scheduling), GERT (Graphical Evaluation and Review Technique), and, more recently, PDM (Precedence Diagramming Method). Those that have surfaced as the more popular methods are CPM (or ADM), the conventional approach, and PDM, favored by software manufacturers.

These diagramming methods are replacing the **bar chart** (known also as the **Gantt chart**), which has been the traditional approach to combining both the planning and scheduling phases. For more effective control of projects, the project management approach, using a network planning method, separates the planning function from the scheduling function. The planning function results in the network diagram (more commonly called the planning diagram), and scheduling, based on the planning diagram, is a manual or computerized calculation process.

PROJECT MANAGEMENT FUNCTIONS

Since 1981, project management has established itself as an independent profession primarily through the efforts of the Project Management Institute (PMI). The institute has developed and supported a comprehensive Project Management Professional (PMP) certification program that requires a qualified candidate to be proficient in certain attributes that include adhering to a code of ethics, acquiring adequate job experience, and possessing specialized knowledge known as the Project Management Body of Knowledge (PMBOK). PMBOK consists of nine management functions: scope, cost, time, human resources, communications, quality, contract/procurement, risk, and project integration. These are defined as follows:

- **Managing the scope** of the project is controlling the project through the aims, goals, and objectives of its sponsor.
- **Managing costs**, required for financial control of the project, is accomplished through accumulating, organizing, and analyzing data and reporting the cost information.
- **Managing time** is planning, scheduling, and controlling the project to achieve the time objectives. (Time management and cost management may form the very substance of project management. These two functions make up most of the contents of this book.)

- **Managing human resources** involves directing and coordinating the administration of people involved in the project.
- **Managing communications** keeps information flowing among members of the project team and management, helping ensure a successful conclusion of the project.
- **Managing the quality** is the fulfillment of the quality standards set up for performance of the project.
- **Managing contract/procurement** includes selecting, negotiating, and awarding orders and administering procurement of material equipment and services.
- **Managing risk** is dealing with the degree of uncertainty of the project through knowledge of and experience with the conditions.
- **Managing project integration** ensures that the various functions of the project are properly coordinated.

Persons just beginning to apply project management need only be involved with the functions that are mostly used: managing time, cost, and scope.

PROJECT PLANNING AND SCHEDULING

The most commonly used project management function is managing time, of which planning is the major component.

The project planning technique divides a project into groups of interrelated **jobs** (sometimes called **work items** or **activities**) headed toward a common goal. Project planning is replacing the use of bar charts because of its superiority over bar charts when planning projects. Project planning can show the interdependencies of jobs, highlight critical jobs, and provide valuable insight into maintaining excellent project performance. Bar charts are ideal for schedule displays but depend on the project plans for the data.

Bar charts had been used (and are still used in some areas) since 1917 in business and industry for planning and scheduling projects. An overview of the use of bar charts is shown in Figure 1.1, which depicts graphically a project's activities and their timing. This type of bar graph indicates the beginning and end dates for each of Jobs A, B, C, and D. These jobs (or activities) represent the major divisions: (A) engineering, (B) material procurement, (C) construction, and (D) equipment delivery and installation, respectively, of a design/build project. In many cases, to better explain a project's planning and scheduling, these major divisions need to be subdivided into additional jobs, detailed on a second and a third chart. If further subdivisions are needed to understand the planning and scheduling process, then additional charts are constructed. For example, the planning and scheduling of engineering would be divided into several components, including structural engineering, electrical engineering, and mechanical engineering. Each of these could become a series of bar graphs, and each could have additional components, with their own individual bar graphs. With a relatively large project, the numerous bar graphs could undermine control of the project.

FIGURE 1.1   Bar chart for planning and scheduling a project

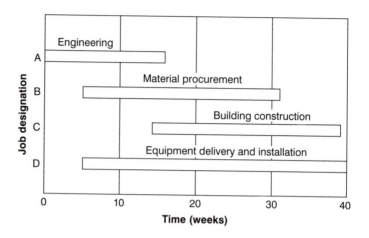

While a bar chart does provide essential scheduling information, it has a segregated and detached effect with all of its accessory charts. Moreover, it does not contain essential planning information, such as job interrelationships among the engineering, procurement, construction, equipment delivery, and installation divisions and their components. This type of chart does not answer the questions that a thorough plan must be ready for, such as:

- What parts of these jobs can be performed concurrently?
- What parts of these jobs must be completed before other jobs begin?
- Must certain jobs or parts of jobs be given priority to avoid holding up the scheduled project completion?
- Do some jobs or parts of jobs have optional starting and ending dates, and what is the extent of these options?

These and many more questions important to setting a project in motion can be answered when generating the project management planning process and, subsequently, the project planning diagram (Figure 1.2). Developing this diagram will require additional information, thought, and effort, but the results are worth it. Chapter 2 explains the necessary guidelines to prepare such a diagram. Also in Chapter 2 is the use of the Work Breakdown Structure (WBS), an important development in producing the project planning diagram. This technique breaks the work activities of the project into smaller, discrete, organized components.

Calculating the project schedule begins once the project planning process is completed. Combining planning with scheduling is not recommended. Any attempt to do this will compromise the true intent of project management. The scheduling procedure can be done with comparatively basic mathematical calculations if the project is small, or with selective computer software designed especially for scheduling projects. Chapter 3 demonstrates the routines associated with scheduling that use the project planning diagram developed in Chapter 2.

Notwithstanding the limitations of bar charts in the planning process, they have been refined over the years to become an excellent way to display schedules. Chapter 3 demonstrates the construction of the bar chart and how it

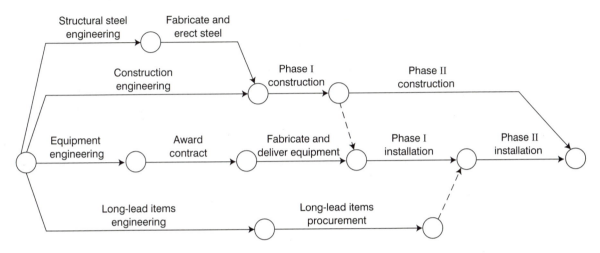

FIGURE 1.2   Project planning diagram

facilitates scheduling. The bar chart also provides a superb communication aid in controlling projects by illustrating their status. Chapter 4 describes the effectiveness of the bar chart in project control.

PROJECT MANAGEMENT APPLICATIONS

The practice of project management principles has grown in industry, business, and government because of its proven merit in helping resolve problems encountered from the ever-increasing complexities of modern-day projects.

Evidence of how extensively project management is used are the ever-increasing numbers who participate in the annual Project Management Institute Seminar/Symposium. People from all over the world present project management papers on various subjects in such areas as construction, utilities, automotive engineering, pharmaceuticals, manufacturing, architectural/engineering, banking, marketing, medical care, and software development.

Because of constant changes in technology and in the marketplace, we cannot rely on traditional organizational structures to respond rapidly to changing conditions. New dynamic methods have emerged. Notable among these is the emphasis on incorporating quality improvement systems in business and industry. Several new management styles based on total quality have dramatically improved efficiencies. Project management principles have enhanced their applications.

Included in the total quality philosophy are the practices of participative management and employee involvement (PM/EI). **Participative management (PM)** is defined as the combination of techniques and skills used to provide employees at all levels with opportunities to participate actively in key managerial processes affecting job-related matters. **Employee involvement (EI)** is defined as the process by which employees at all levels have the opportunity to participate actively in key managerial processes affecting job-related matters.

These groups are often called **quality circles**, and they all have closely related responsibilities—to participate in almost every facet of operations including goal setting (similar to setting objectives, an initial step in project management), planning, organizing (analogous to developing the work breakdown structure), and decision making. The universal goal of PM/EI or quality circles is to develop a management style that accentuates a participative, cooperative relationship among all personnel levels of an organization to fulfill certain objectives.

One of the predominant guiding principles included in Ford Motor Company's Statement of Mission, Values and Guiding Principles embodies PM/EI: "Employee involvement is our way of life. We must treat each other with trust and respect."

Ford Mustang Program

A fine example of applying project management principles to assist in achieving success was Ford's 1994 Mustang model program. This program embodied World Class Timing (WCT), an innovative product development process that brought the Mustang to market faster, more efficiently, and with higher quality. The WCT process consisted of a framework of detailed plans and schedules that blends in every process used in developing the Mustang. Twenty disciplines were involved, including design, engineering, manufacturing, purchasing, marketing, personnel, and finance.

The major objective Ford's senior management set for the Mustang program was to improve customer satisfaction. Close behind were mandates to cut 20 percent from normal product development costs and 25 percent from the 48-month concept-to-customer cycle.

The Ford Mustang team actually surpassed the objectives. This $700-million project reflects a cut of nearly 30 percent, and the first Mustang came off the assembly line for delivery just 35 months after senior management authorized its development—one month earlier than the stated objective. In addition, the program was completed 13 months earlier than most of the other prevailing model programs were scheduled to be completed.

A great deal of the success can be credited to intensive early planning. It all started with Ford's car programs office laying out 150 major work activities to a team placed "under one roof." The team members came from design, product development, engineering, manufacturing, purchasing, suppliers, and assembly, and all were involved early. The main expedient in orienting this multidisciplinary team was the project plan. It enabled execution of many work activities simultaneously instead of sequentially, the norm in traditional product development work. Time was saved by being able early in the program to find the problems and to make the necessary corrections. Additional costs were minimized by avoiding "panic" situations later in the program, which would require excessive overtime, material corrections, and design changes.

WCT reduces the concept-to-customer time from four to three years, which allows transfer of facilities, personnel, and brainpower to another program one year earlier. Many of the 450-member Mustang team moved to the new Taurus team.

Ford's goal is to reduce the product cycle time further by enhancing the present procedures, including the project management procedures. Reducing WCT from 36 months to 30 months is doable; Ford expects to find a way.

Hospitality Industry

Major hotel and restaurant companies are moving to adopt total quality management goals for making service more efficient and dining more pleasant to the customer. One of the doctrines of total quality management is to transfer responsibility for making decisions down as far as possible on the chain of command. For the hospitality industry, this meant forming quality circles of people who work directly with the customer. This group can best determine what works best to keep guests satisfied. The quality circle concept empowers its members to come up with ideas and the means to set them in motion.

Members of one team made up of servers at a Ritz-Carlton hotel in South Florida prepared a planning diagram of the activities necessary to serve breakfast to guests. This resulted in reducing significantly the amount of time that it had normally taken for guests to be served.

With some training in the preparation of planning diagrams, the employees are using this technique, along with other project management methods, to take care of various problems. Complaints have dropped as much as 70 percent at one location since staffers have set up quality circles.

Federal Programs

In 1993, Congress passed the Government Performance and Results Act, requiring each major agency to submit a strategic plan for its programs. Each plan must contain

- the major functions and goals of the plan
- the personnel, capital, and other resources required to meet the goals and objectives
- performance goals of each plan, and how they are related to meet the general goals and objectives of the strategic plan
- external factors that could affect achievement
- program evaluations used in establishing or revising goals and objectives

The goals and objectives are to be unbiased, quantifiable, and measurable. Under the provisions of this Act, all of the above need to be in place by September 1997, with plans to cover a period of five years. Program performance reports will be starting in year 2000. All federal agencies understand that for all of this to be realized, they may need to adopt project management principles for many of their programs.

The federal government is no stranger to project management applications. After all, it was the U.S. Navy in the late 1950s that pioneered the network diagramming approach for planning, scheduling, and controlling projects. In the ensuing years, government specifications required bidders to prepare project planning diagrams and computerized schedules for construction of their facilities and related work. Besides the Navy, federal agencies requiring these project management standards include the U.S. Army, U.S. Air Force, Environmental Protection Agency, and the National Aeronautical and Space Agency.

Government bureaus require extensive documentation for the work that is to be done, and project planning diagramming is useful in helping satisfy this requirement. Project planning diagramming is also being used for administrative and systems programs as well.

BENEFITS AND LIMITATIONS OF PROJECT MANAGEMENT

Using project management techniques to guide and control the course of a project affords more confidence in a successful completion. Value judgments are still necessary, and the good results are still the product of dedicated effort. Most important to the success of project management techniques is the complete support of all levels of management, especially top management. They must understand and accept its underlying principles. Although project management requires effort, participation, dedication, and cooperation, the resulting rewards make it all worthwhile. Advantages include the following:

- A clear picture of the scope of the project forms that is readily communicable. It clearly shows responsibilities of those participating in the project.
- It is a credible aid to familiarize new personnel on the details of the project.
- It forces those participating in the project to think it through in greater detail.
- It provides a means of accountability by assigning those responsible for the specific activities that make up the project.
- It is an excellent way to define relationships among the major areas of responsibility as well as among individual project activities.
- It enables early and prompt corrections to the project to meet changing or unpredictable conditions. It evaluates strategies and objectives with the help of the project planning diagram and the computerized schedule.
- It advances the professional development of the project team members through training and active participation.

Several limitations need to be addressed and overcome to allow the benefits to be realized:

Limitations of Proj. Mgmt.

- The project plan cannot be complicated. It needs a certain degree of simplification to avoid any misinterpretation.
- Changes to the project planning diagram may require more replanning time than can be tolerated.

SUMMARY

A project consists of a series of activities arranged in a logical order. Projects have well-defined objectives and achieve a specific product or service. Projects exist in most fields of endeavor, including designing, developing, and introducing a new product; installing and/or relocating manufacturing equipment or a process; designing and constructing a manufacturing or commercial building; relocating offices; and introducing a new business system or software product.

Project management is defined as managing and directing time, costs, personnel, and other resources to complete a project in an orderly, efficient, and economical manner while meeting the timing objectives, staying within budget, and obtaining quality results. Project management is knowing where you are in a project and managing changes when problems arise to achieve successful project goals.

Project management techniques are applied during three stages in a project: planning deals with what has to be done, scheduling concerns when the work needs to be done, and controlling monitors and measures the performance of the project.

Project diagramming methods that are applied in the planning stage have replaced the conventional bar chart as the principal planning tool. The two preferred methods currently used are the critical path method (CPM)[1] and precedence diagramming method (PDM). Bar charts continue to be constructed for displaying project schedules.

Based on the project plan, the project and its activities are scheduled either manually (limited to small projects) or with the computer. The many available software programs, once understood and found compatible, are invaluable for preparing of schedules and project control.

Project management has established itself as an independent profession. A person successfully completing the certification program is qualified as a Project Management Professional (PMP). The PMP is required to be proficient in the nine management functions known as the Project Management Body of Knowledge (PMBOK). PMBOK's functions are scope, cost, time, human resources, communications, quality, contract/procurement, risk, and project integration. (This book is essentially dedicated to the cost and time functions, which are the very essence of project management.)

---

[1]Arrow diagramming method (ADM) and activity-on-arrow refer to the same critical path method and will be used interchangeably throughout the text.

The operating standards of many businesses, industries, and governmental agencies incorporate project management functions. Specifically, fields practicing project management include construction, utilities, automotive industry, pharmaceuticals, architecture, engineering, banking, marketing, manufacturing, medical care, and software development.

Using project management methods to guide and control the project offers high confidence of success. It does require more effort because there is more detail to consider, but the discipline will make the users think through the project.

This book provides basic knowledge for those wishing to become familiar with project management and the ability to apply the technique to their own projects. By understanding the basics, the student should be able to apply the techniques to more advanced projects.

Exercises

1. Define a project.
2. Name some projects in your general geographic area that have been completed recently or are now under way. Identify at least four distinguishing characteristics that show it is indeed a project.
3. Define project management. List the three stages, with a brief description of each stage.
4. Name the two project diagramming techniques initiated in the 1950s that pioneered project management development. Name a third diagramming technique that has increased in popularity in recent years.
5. A certified Project Management Professional (PMP) must be proficient in the nine Project Management Body of Knowledge (PMBOK) functions. Name two of these functions that form the very essence of project management, and describe each briefly.
6. Proponents of project management list many advantages to its use for attaining project success. Name at least four advantages and one possible disadvantage.

The Project Plan

THIS CHAPTER WILL COVER

- Producing a project plan

- Developing a work breakdown structure

- Constructing a planning diagram
  a. Using the arrow diagramming method
  b. Using the precedence diagramming method

Planning involves organizing a project in a logical order, identifying and defining the work activities in such a manner that they will help achieve the project objectives. Many authorities consider project planning the most important function of project management. The history of project management confirms this, since the first function developed was a disciplined approach to planning. When specific techniques recommended for planning were first introduced, they met with skepticism. The author encountered resistance when he first introduced a new planning method on a facility project at Ford Motor Company. A typical comment came from one of the managers who was to be in charge of a major portion of the project. This manager, unaccustomed to any "paperwork" planning, remarked: "I always do that stuff in my head."

The planning process that is represented by a planning diagram is a graphic display of the **logical thought** mechanism used in developing the project plan. The advantage of using the planning process is that it allows more than one person's contribution to its development. The graphic display provides an image of the project scope. Preparing this image by exercising project management methods compels all who are participating to think the project through.

Those who have the experience and knowledge associated with the project should prepare the project plan. They have the assignment to prepare the **what has to be done** list. The planning phase, unlike any of the other phases, relies more heavily on the human thought process in developing the logic, with minimal help from any computer software. The computer comes into play when developing the schedule and for project control.

THE PLANNING PROCESS: AN OVERVIEW

Planning a project begins with preparation of comprehensive statements of the objectives, usually determined by management directives. These objectives usually include (1) when the project will be completed, (2) the costs budgeted for this project, and (3) the expected results.

Once the objectives are set, those assigned to the project begin to identify the activities (or jobs) necessary to complete the project. A project management technique, work breakdown structure (WBS), becomes a useful tool in organizing this work. The final step is diagramming the project plan in a flow pattern that arranges the jobs in a logical sequence, recording relationships to provide a clear "picture" of the project.

Two diagramming methods are used for project planning: **precedence diagramming method (PDM)**, sometimes referred to as the **activity-on-node** method; and the **arrow diagramming method (ADM)**, sometimes referred to as the **activity-on-arrow** or the **i,j** method.

ADM is the project planning diagramming procedure this book employs for a step-by-step planning approach, as it tends to produce an easily understood plan. This diagramming method has other advantages, the dominant one being the ability to depict the project organization to indicate who is responsible and

ADM: enables ability to depict who is responsible for each project

accountable for each project activity. The project planning diagram clearly shows the major divisions of the project and their interrelationships.

PDM has many supporters that prefer its style. So that the reader can weigh both diagramming methods, both will be demonstrated in planning the sample project.

The intent of both of these diagramming methods is the same: to communicate what is to be done in the project to all involved persons.

The mechanics of drawing a proper CPM diagram require working knowledge of such terms as arrows, dummy arrows, nodes, and activities. An arrow is drawn for each activity. The sequence of the arrows indicates the flow of the work from the beginning to the end of each project. Generally, the complete project diagram will have a beginning point and an ending point. (Some projects may have several beginning and ending points.) Arrow junctions are called nodes (or events). Nodes are numbered, not necessarily in any sequence, to identify the location of each activity in the project planning diagram.

The step-by-step approach to project planning is best illustrated by going through an actual project. A design/build project has been selected, since it includes the major activities associated with the various engineering design disciplines, equipment procurement, and building construction to complete the project.

THE PLANNING PHASE

Planning can be the most time-consuming phase of the total project; however, the time spent can also be the most rewarding. Those who have completed the planning phase can be confident that they have a complete knowledge of what the project entails.

Planning a project will follow these steps:

1. Establish objectives.
   a. State the objectives (e.g., project start and finish dates, budgets, technical results). These will reflect management directives that motivated the project.
   b. List interim objectives (or **milestones**). These are significant events within the time frame of the project that must be realized to meet the management objectives.
   c. Designate responsible personnel or departments needed to meet the objectives. These are the groups whose participation is essential to successful execution of the plan. They need to be selected early in the project planning stage and should be identified when setting the objectives.
2. Develop a plan.
   a. List the jobs (or activities) that must be done to complete the project.
   b. Develop the WBS. Prior to developing the WBS, group the jobs according to the nature of their work and under departments responsible for them.

   c. Delineate the jobs by determining their relationships.
   d. Determine which jobs precede every other job and those that succeed every other job.
   e. Determine which jobs can be done concurrently.
3. Construct the project planning diagram.
   a. Show the sequence in a planning diagram. The planning process has reached an initial completion stage when a graphic display of the project work items (jobs) and their interrelationships can be completed.
4. Identify the timing duration of each activity on the planning diagram.
5. Identify the costs and labor/personnel associated with each activity on the planning diagram. (While displaying personnel/labor and costs on the planning diagram allows for a more complete understanding of the project, detailed discussions of these two very important items will be deferred until they are discussed in detail in later chapters.)

THE DESIGN/BUILD PROJECT

Snyder Engineering Associates, a full-service engineering and construction company, has been selected to design and build a manufacturing building for a client, Christopher Development Company. This building will accommodate a fabrication area, office space, and maintenance and storage areas. In this building Christopher Development plans to fabricate and assemble household ventilating fans. Christopher wants this project to start no later than October 2, 1995, and be ready for occupancy by April 22, 1996, a total duration of 29 weeks. Christopher's board of directors has authorized a budget of $2.4 million for this project.

From past experience Snyder Engineering realizes that the 29-week deadline can only be met by working an accelerated seven-day-a-week schedule. Furthermore, to use the time most effectively, Snyder needs to exercise selected project management techniques to prepare an effective plan and schedule so that this project can be completed on time and within budget. (The plan and schedule that would apply these project management methods should also confirm the earlier decision by Snyder Engineering on the need for a seven-day workweek schedule for this project.)

When Snyder Engineering first worked with Christopher Development in preparing overall objectives, Snyder appointed Bob Sachs to lead this project. This was arguably a good choice. In addition to past experience, technical knowledge, and expertise in managing design/build projects, he also has the attributes necessary for a successful project manager—leadership, communication skills, planning and organization skills, and interpersonal and team-building skills. Sachs will be responsible for selecting the team members who will assist in preparing the project plan. These individuals should also have the experience to handle their assignments and the appropriate attributes for their responsibilities.

In addition to developing the project plan and schedule, the team members will be charged with (1) generating a cost budget schedule for proper and timely disbursement of funds, and (2) allocating resources, including personnel, skilled labor, material, and equipment. They will also be responsible for setting up a communication system that will allow periodic monitoring of the project's status to ensure its successful completion.

Planning the Christopher Design/Build Project

Planning the Christopher Design/Build Project will follow six procedural steps: establishing objectives, defining the required activities, dividing those activities, constructing the WBS, drawing subdiagrams, and developing the project planning diagram.

Step 1: Establish objectives. Usually set by higher management, these are the major objectives: project start and completion dates; project cost budgets; and the expected product or results. The timing objectives may also contain interim timing objectives (generally known as milestones) that must be met to achieve the completion date. (Figure 2.1 shows a display of the main objectives, milestones, and the sections responsible for completing these objectives.)

Step 2: Define the required activities. Once the objectives are authenticated, the next step is to identify the jobs or activities needed to complete the project. Identifying the activities is done by those persons and/or divisions noted on the objectives and others who have the most knowledge and experience of the work to be done. This step is quite an effort to complete—it is essentially a checklist of the work to be done—and can be time-consuming if it is to be thorough. For com-

Objectives/Milestones	Date	Responsibility
Start Christopher Design/Build Project	October 2, 1995	Project Manager (Bob Sachs)
Complete building design; start phase I building construction	January 22, 1996	Engineering Manager (TBA) Construction Manager (TBA)
Start mechanical and electrical equipment installation	February 12, 1996	Engineering Manager (TBA) Construction Manager (TBA)
Start phase II building construction	March 25, 1996	Engineering Manager (TBA) Construction Manager (TBA)
Complete Christopher Design/Build Project	April 22, 1996	Project Manager (Bob Sachs)

FIGURE 2.1 Objectives for the Christopher Design/Build Project

Activity	Description
1. Design structural steel	Design and detail the structural steel based upon the building design and the owner's requirements.
2. Fabricate and erect structural steel	Fabricate structural steel based upon the design; erect steel when fabrication is complete.
3. Design building	Upon review of owner's requirements, design building concurrent with structural steel design.
4. Design mechanical equipment	Upon review of owner's requirements, design appropriate mechanical equipment.
5. Construct phase I	After building design and structural steel are completed, construct the building "shell": roof, siding, floor slab.
6. Construct phase II	After the building "shell" is "closed in," start phase II: installation of the building mechanical, electrical, and interior items.
7. Fabricate mechanical equipment	Upon completing design of the mechanical equipment for the building, solicit bids, negotiate, and award contracts for best proposals to fabricate and deliver mechanical equipment.
8. Install mechanical and electrical equipment	Start installation of mechanical and electrical equipment when phase I is complete and mechanical and electrical equipment is on site.
9. Design electrical equipment	When design of building and mechanical equipment items is essentially complete, start design of electrical equipment, power requirements, lighting, and electrical accessories for mechanical equipment.
10. Procure electrical equipment	Upon completing design of the electrical equipment, solicit bids, negotiate, and award contracts for best proposals to fabricate and deliver electrical equipment.
11. Design interior items	Architectural designs for interior work—building office partitions, window and door treatments, office equipment, etc.—can begin concurrent with building design and after mechanical equipment designs are essentially complete.
12. Procure interior items	Upon completing interior item designs, solicit bids, negotiate, and award contracts for best proposals to manufacture and deliver interior items.
13. Install interior items	When interior items are delivered, install concurrent with phase II construction.

FIGURE 2.2   Activities (or work tasks) for the Christopher Design/Build Project

plete particulars, the project team prepares a work description for each activity to allow for a more complete understanding of the project. Description of each activity for the Christopher Design/Build Project is shown in Figure 2.2.

Step 3: Divide activities into work groups.   To assist in determining relationships between activities, it is desirable to first group the closely related functions under major categories or divisions. For the Christopher Design/Build

FIGURE 2.3 Groups
representing major sections
of work for the Christopher
Design/Build Project

Major Sections of Work
_____

Structural
- Design structural steel
- Fabricate and erect structural steel

Building
- Design building
- Construct phase I building
- Construct phase II building

Mechanical
- Design mechanical equipment
- Fabricate mechanical equipment
- Install mechanical and electrical equipment

Electrical
- Design electrical equipment
- Procure electrical equipment

Architectural
- Design interior items
- Procure interior items
- Install interior items

Project they are structural, building, mechanical, electrical, and architectural. Setting them up in logical groupings simplifies the task of determining how the jobs relate to each other. (See Figure 2.3.)

Step 4: Construct the WBS. The WBS amplifies Step 3 by graphically setting up the groupings in the form of an organization chart. Listed under each group are the work assignments (or activities) needed to complete this specific group. Persons (or departments) responsible for each major assignment can also be shown (as an option) on the WBS chart. (The WBS for the Christopher Design/Build Project is shown in Figure 2.4.) Using WBS charts as a prelude to preparing the project planning diagram has become increasingly popular, and it is recommended that this form of chart be prepared for project planning. If necessary, replace the tabulation method with the WBS approach once its construction is fully understood.

Step 5: Draw subdiagrams. To further assist in the development of the project planning diagram, the subdiagramming style sets up sequentially preliminary diagrams to show relationships of major divisions as well as individual activities. The first steps of the subdiagrams (Figure 2.5) are done by the individual departments or areas of responsibility to show the delineation of their work activities within their own department. The final step of the subdiagrams (Figure 2.6) shows the interrelationships of the work among the areas of responsibility. If the project is extensive, a series of meetings may be necessary to resolve these relationships.

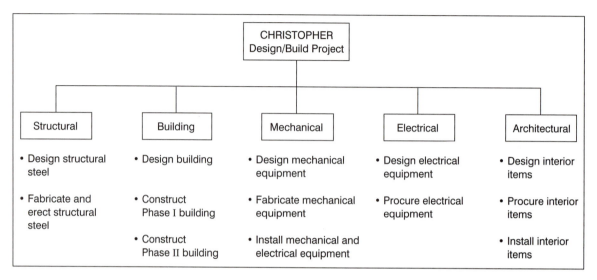

FIGURE 2.4   Work breakdown structure (WBS) for the Christopher Design/Build Project

Thoroughly prepared subdiagrams will become a valuable aid to reducing the work involved in preparing the final step, the project planning diagram. Once the subdiagrams are completed, the project team will have reached a fairly good understanding of the entire project.

Constructing the project planning diagram with the arrow diagramming method requires an understanding of arrow diagramming rules.

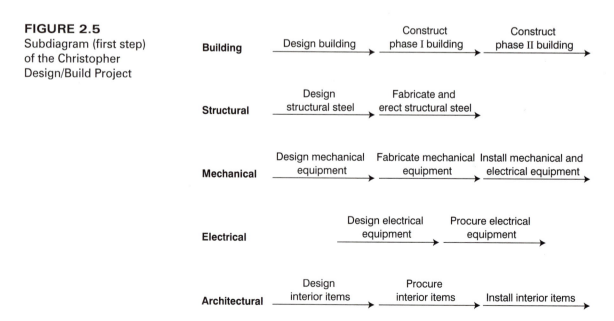

FIGURE 2.5
Subdiagram (first step) of the Christopher Design/Build Project

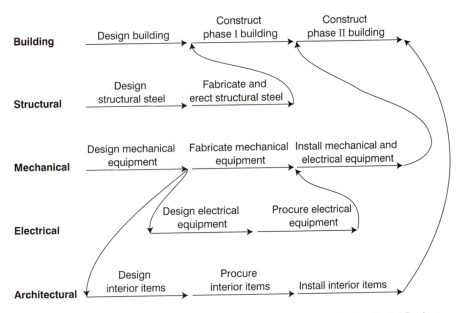

FIGURE 2.6 Subdiagram (final step) of the Christopher Design/Build Project

Arrow Diagramming Rules. The basis of a project planning diagram (or network diagram) is a graphic diagram of arrows representing specific activities (or jobs). The manner in which these arrows are connected depicts the interrelationships among these activities.

In an arrow diagram, an arrow is used to represent an activity.

The work is assumed to flow in the direction in which the arrow points. It begins at the left end of the arrow and finishes at the right end or head of the arrow. The arrows are not time-scaled. (If time-scaled, projects of long duration containing individual activities of long duration would be unmanageable to construct.) Sequences of activities are indicated by the way the arrows are interconnected:

1. *Design structural steel* must be completed before *Fabricate and erect structural steel* is started.

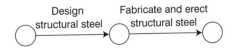

2. *Design building* can be done concurrently with *Design mechanical equipment.*

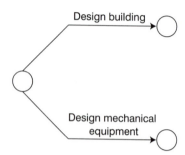

3. *Fabricate and erect structural steel* and *Design building* can be done concurrently and must be completed before *Construct phase I building.*

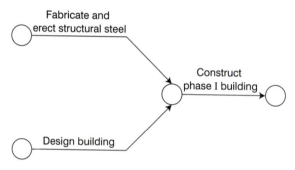

Two types of arrows are used in diagramming:

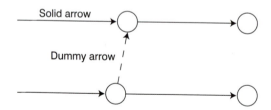

1. **Solid** arrows represent jobs or activities.
   a. The solid arrow represents a job or activity that consumes time (has a duration) and resources.
   b. Work is assumed to flow in the direction in which the arrow points.
   c. The length can vary (time or duration is not symbolized by the length).
2. **Dummy** arrows show special interrelationships.
   a. Dummy arrows are drawn as dashed lines.
   b. Dummy arrows do not represent a job or activity; they have no duration and do not consume any resources.

    c. When drawing the diagram, dummy arrows are used as an expedient to separate and identify the major divisions of the project.

    d. Dummy arrows are used at times when there may be confusion in identifying activities.

The beginning and ending of each arrow is called a **node**. A node has the same credential as an event. Nodes are noted as circles on the diagram and are numbered for identification, but are not necessarily numbered in sequence.

    1. The circles with numbers 1, 2, 3, 7, and 11 are nodes.

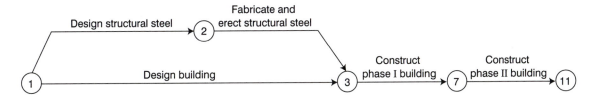

    2. A node or event is a point in time and has no duration. For example, Node 1 is an event, start *Building design;* Node 3, complete *Building design* (and can also be termed start *Phase I construction*).

    3. Activities use their start and complete nodes for identification. For example, *Design building* is identified as activity 1,3; *Construct phase I,* activity 3,7; *Construct phase II,* activity 7,11.

    4. Each arrow (job or activity) is identified by two nodes: the **i node** is at the beginning of the arrow, the **j node** at the end of the arrow (i,j node references are used at times to identify the activity in CPM). *Critical path method*

Effective planning of the jobs or activities in a project requires adherence to arrow diagramming rules in constructing the project planning diagram.

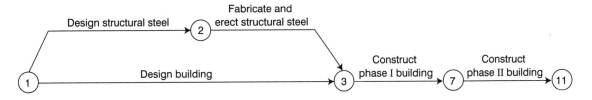

    1. Activity 2,3, *Fabricate and erect structural steel*, has the following positions in the project:

        a. *Fabricate and erect structural steel* follows *Design structural steel.*

        b. *Fabricate and erect structural steel* precedes (is completed before) *Construct phase I building.*

        c. *Fabricate and erect structural steel* is performed concurrently with *Design building.*

        d. *Fabricate and erect structural steel* and *Design building* need to be completed before *Construct phase I building* can start.

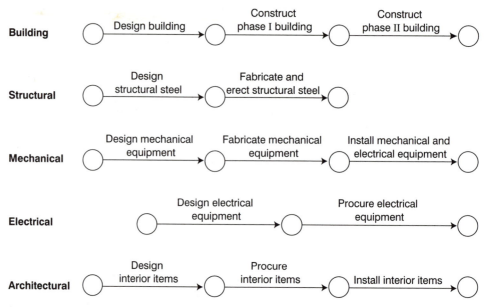

FIGURE 2.7 Planning diagram (first step) of the Christopher Design/Build Project

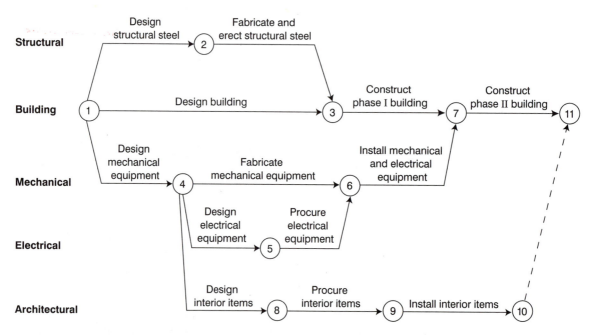

FIGURE 2.8 Planning diagram (final step) of the Christopher Design/Build Project

2. When two or more arrows (activities) end at the same node, this does not necessarily mean that the work they represent will be scheduled to start at the same time. (Chapter 3 has a detailed explanation of this scheduling situation.)
3. Similarly, two or more arrows that end at the same node do not necessarily have an identical finish date.

Step 6. Develop the project planning (arrow) diagram.   Finally, the project planning diagram is developed showing the relationships of all of the activities (Figures 2.7 and 2.8). If the intermediate steps are followed (very important when planning large and/or complex projects), this step is essentially a refinement of the final step of the subdiagrams. By adding the node symbols and the numbering identification, the basic project planning diagram for scheduling purposes is essentially complete.

PRECEDENCE DIAGRAMMING   *critical path method*

In addition to the conventional CPM approach for diagramming a project plan, the precedence diagramming method (PDM) is worthwhile to consider. In fact, PDM has gained considerable popularity and acceptance among project management software manufacturers and establishments that are engaged in very large, complex projects. The software manufacturers prefer PDM because it is easier to program and they can more easily arrange graphics on the computer monitor screen with PDM diagramming. Extensive details that may be necessary for large and complex projects can be handled more easily with PDM diagramming.

While using PDM is thought to avoid certain complexities in drawing the project planning diagram, questions remain regarding its ability to fully comprehend the project. PDM has certain peculiar characteristics and their implications that the project manager, those analyzing the project, and others associated with the project must deal with. For the reader to understand how PDM is employed, let us first explain briefly the symbols used, then construct a PDM diagram for the Christopher Design/Build Project. (Keep in mind one major difference between CPM and PDM: In CPM the description of the project activities are shown over the arrows; PDM, within the nodes.)

The underlying objective of any diagramming technique is to portray the project plan graphically so that it can be used effectively. To be adept at PDM, one needs to master its symbology (see Figure 2.9).

PDM has four precedence relationships (CPM has one): **FS** or Finish to Start (the one that is common with CPM); **FF**, Finish to Finish; **SS**, Start to Start; **SF**, Start to Finish. There are time considerations known as the **lead-lag** factors:

1. **Lead** is the amount of time by which one activity precedes the following activity.
2. **Lag** is the amount of time by which one activity follows the preceding activity. (**Lead-lag** factors avoid the need to use dummy arrows.)

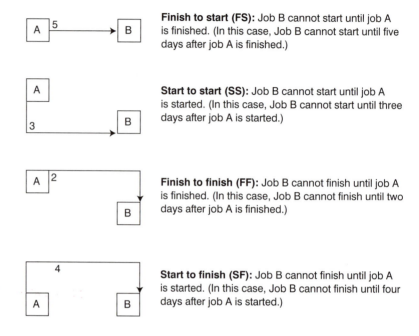

FIGURE 2.9 Symbols used to show precedence diagramming relationships.

Finish to start (FS): Job B cannot start until job A is finished. (In this case, Job B cannot start until five days after job A is finished.)

Start to start (SS): Job B cannot start until job A is started. (In this case, Job B cannot start until three days after job A is started.)

Finish to finish (FF): Job B cannot finish until job A is finished. (In this case, Job B cannot finish until two days after job A is finished.)

Start to finish (SF): Job B cannot finish until job A is started. (In this case, Job B cannot finish until four days after job A is started.)

Constructing the project planning diagram still requires the initial steps previously discussed in this chapter as employed in the CPM technique—setting the objectives, listing the project activities, and preparing the WBS. Instead of using the subdiagrams approach for arrow diagramming, PDM diagrammers will tabulate the successors and predecessors of each activity before proceeding with the PDM construction of the Christopher Design/Build Project. (See Figure 2.10 for a finished PDM version of the Christopher Design/Build Project.)

As stated earlier, the CPM approach will be used for the purposes of this book. Those desiring to use PDM should prepare their diagram at this time before proceeding. While not obvious to advanced PDM users, beginners may find it difficult to use in the basic project planning and scheduling procedures. The time and effort spent to understand the PDM symbology and the mechanics of lead-lag links are too much of a challenge in this text, whose immediate purpose is to prepare the reader (or student) in a minimum time frame to prepare a project plan.

There are other reasons to favor CPM at this time: Once the PDM planning diagram is completed, the peculiar characteristics of a PDM diagram make it somewhat difficult to analyze except for the advanced user. As we proceed from the planning phase to the scheduling phase and interpreting the effect of **critical project activities**, PDM users must take in consideration many more variables. Recently, several leading authorities have disclaimed the need for lead-lag links, except in a few situations. Without these, PDM is more similar to CPM.

Inputting PDM data into the computer program appears to be more complex and requires additional time because of the FS, FF, SS, SF, Lead, and Lag data that need to be entered. (See Chapter 7 for the PDM input for the Christopher Design/Build Project.)

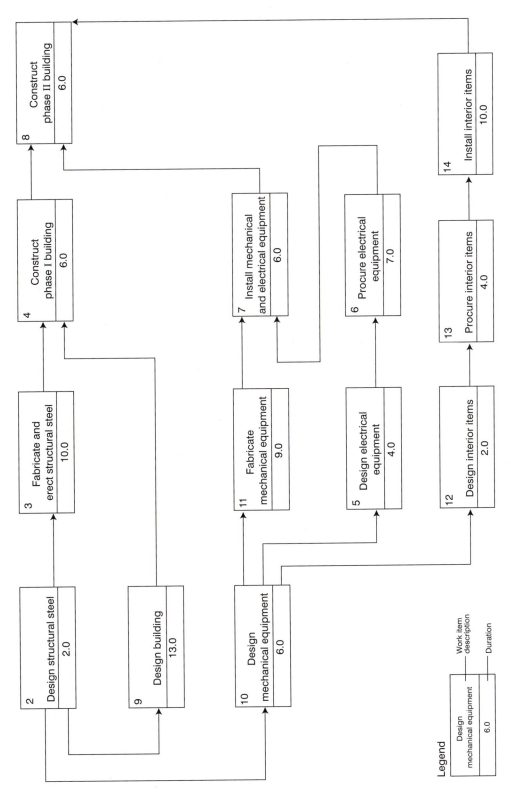

FIGURE 2.10 Precedence diagram for the Christopher Design/Build Project

Legend

	Work item description
Design mechanical equipment	
6.0	Duration

SUMMARY

Planning what needs to be done is the first phase of the project management process and the most important in bringing a job to a successful completion. In our complex society, we can no longer rely on hunch and intuition as the basis of decisions. Good judgment is still necessary, and together with the use of the planning discipline, reliable decisions can be made.

A disciplined approach to proper planning follows these sequential steps:

- Establish objectives and milestones. Objectives are usually established by upper management.
- Identify the project activities (or jobs), preparing a checklist of the work to be done.
- Construct the work breakdown structure (WBS), an organized arrangement of the work to be done, showing the major divisions and the work associated with each division.
- Delineate the project activities, showing interrelationships of all project activities and the major divisions involved in the work.
- Construct the project planning diagram, a graphic picture of the total project

The two most popular project planning diagramming methods are the critical path method (CPM) and the precedence diagramming method (PDM). Specific rules apply to use of both these methods. Proponents of each method can enthusiastically argue why one should be chosen over the other, yet they both achieve the same result. This book uses the CPM method for most applications because it is easier to explain and simpler to use.

The complete project planning process, concluding with the project planning diagram, will include for each activity the following: time duration, costs, labor/personnel requirements, and responsible person or department. Milestones (another name for interim objectives) can also be shown on the diagram. (All of these items will be discussed in more detail in appropriate chapters, then incorporated into the project planning diagram.)

Developing the project planning diagram forces the team to think through a project. Although this approach is time-consuming and taxes the intellect of those participating in its preparation, it has rewards. Early efforts will become especially valuable when problems arise and the project planning diagram helps the team make intelligent responses and solutions.

Exercises

1. What are the major steps required to complete the planning phase of the project?
2. An important effort when developing the plan is the WBS.
   a. What is WBS an abbreviation for?
   b. Draw a WBS chart to show these major divisions in the product development of a new alternator: design, develop, test and manufacture.

3. Draw a project planning diagram for the following project:

   A boat is in need of repairs. The project activities needed to repair the boat are as follows:

   - At project start, *Fabricate parts* and *Take boat to repair yard* can begin.
   - *Repair boat* begins after *Take boat to repair yard* has been completed.
   - *Return boat* can start after completion of *Repair boat.*
   - *Fabricate parts* and *Return boat* must be completed before *Install parts* can begin.
   - Project is complete once *Install parts* is complete.

4. Your architectural engineering firm as been chosen to prepare designs and specifications for a building project.
   a. Prepare a WBS for the engineering of the building, whose major divisions are the civil, architectural, electrical, and mechanical engineering departments.
   b. Prepare a WBS for constructing a building whose major work divisions are site grading, foundations, steel framing, roofing and siding, carpentry, and the mechanical and electrical items.
   c. Construct a summary-type planning diagram that consists of just the major work divisions in Exercise 4b. The logical sequence is as follows:
      - *Site grading* starts the construction project.
      - Start *foundations* when *site grading* is completed
      - When *foundations* are completed start *steel framing.*
      - Start *roofing and siding* when *steel framing* are done.
      - Start *carpentry* when *roofing and siding* is complete.
      - Start the *mechanical and electrical items* when the *carpentry* is complete.
      - The project is finished and ready for occupancy when the *mechanical and electrical items* are completed.

5. Prepare the project plan using the precedence diagramming method (PDM) for the project shown in Figure 1.2.

Project Schedule

THIS CHAPTER WILL COVER

- Timing estimates
- (Manual) scheduling procedure
- (Manual) scheduling calculations
- Computer scheduling

Planning relates to *what* work is to be done; scheduling, *when* the work is to be done. The scheduling phase begins after completing the project planning process and constructing the project planning diagram. It is important not to overlap the planning and scheduling phases. Attempting to schedule project activities during the planning process tends to distract the team from the logic functions. Experience has shown that concurrent efforts to plan and schedule cause timing discrepancies and a potential distortion of the project schedule. The project management approach recognizes this dilemma by requiring a completed project plan, with a project planning diagram, to be completed first before the scheduling process can be performed effectively.

This chapter explains how to create the initial timing schedule. A complete schedule will also include the scheduling of resources, costs, and personnel/labor. Chapters 5 and 6 discuss these issues and incorporate them into the sample project. (After completing Chapters 5 and 6, you may elect in future projects to identify personnel/labor and costs at the same time you develop timing estimates.)

Using manual calculations to develop the schedule provides the background and analysis that should be understood before using the computer for calculating the schedule. While computer programs are used sparingly in project planning, they are quite helpful in performing the calculations needed in project scheduling. Scheduling fundamentals are explained and demonstrated in the first part of this chapter, while the second part uses the computer to calculate the project schedule. Following the procedure in this chapter will permit computer scheduling to be an uncomplicated, gratifying task.

MANUAL SCHEDULING: AN OVERVIEW

The manual scheduling method involves an initial analysis of the project prior to devoting time and effort to input the computer data. Making manual timing adjustments prior to involving a computer effort will save time and wasted effort inputting data. Manual calculations require adding and subtracting functions—and accuracy. Initial scheduling can usually be done expeditiously by placing the scheduled start and finish times of each project activity directly on the planning diagram. This chapter will explain and demonstrate how to do this using the sample Christopher Design/Build Project.

The initial calculating effort, **earliest start time**, allows early review of the project duration, a quick check that it meets project objectives. If these are not met, changes can be made early in the planning stage, thereby avoiding wasted work and other problems later as the project evolves. Project team members using computerized scheduling will go through the manual method at this time to evaluate completion timing. (The manual method requires so little time compared to the total computer inputting requirements. The case histories in the later chapters illustrate this approach.)

This chapter also explains how to develop the total float and identify the critical items (which have **0 float**). **Critical items** determine the length of the project. **Total float** of a project activity is the difference between how much time is available to perform that activity and how much time is required. A continuous chain of critical items from the start to the finish of a project, termed the **critical path**, determines the project duration. Float and critical path are an important part of scheduling.

Scheduling using this method has the following principal benefits:

- It establishes a supportable project duration.
- It identifies the critical project activities.
- It identifies those project activities whose scheduled start and finish dates are flexible and can be changed without affecting the project duration. These activities are known to have float.

SCHEDULING PROCEDURE

In the scheduling phase, we are concerned primarily with timing—how much time each job requires to complete and when each job will be scheduled to begin and end. Scheduling a project begins after the sequence of project items has been planned and laid out in a project planning diagram. By using the following procedure, you will be able to schedule each project activity and determine the project duration.

1. Determine the required time (time estimate) to complete each project item.
2. Calculate the available time to complete each project item.
3. Compare the required time with the available time of each project item for its float.
4. Identify the critical (0 float) project items.
5. Determine the float times of the noncritical items.
6. Calculate the duration of the project. If it does not meet the objectives and is not acceptable, make the necessary adjustments to the plan and to the timing estimates.
7. Validate the time schedule by getting concurrence from all of the concerned persons.
8. Prepare a bar chart time schedule.

TIMING ESTIMATES

Estimating the durations of project activities is dynamic; that is, it can be an unending task because of uncertainties that occur throughout the duration of the project. Nevertheless, it is the starting point in the scheduling procedure, whether manual or by computer. Estimating occurs in three main phases:

- initially, when the project planning diagram is essentially complete
- during the scheduling phase, when the initial estimates require revisions because they do not meet objectives

- ongoing estimate changes of selected project activities during the execution of the project to keep it on its planned schedule

The initial estimates are based on how much time the activity should actually take to complete—**required time**—if everything proceeds on a normal basis. Persons familiar with the work to be performed should make these time estimates based on their experience and best judgment. Since they may be held responsible for their estimates, they tend to overstate the normal estimating time to provide for unforeseen conditions. Such cushions may add additional time to a project that may not be tolerable. Excessive use of contingencies may distort the project to the point of making it unworkable. These situations are readily noticed during scheduling, and revised estimates can be called for at that time. (Later in this chapter is a discussion of the three-time estimate approach, which can be used to make reasonable estimates. This method tolerates a moderate use of contingencies.)

With this scheduling process, the initial project duration can be calculated before too much effort is expended. If the time does not meet objectives, adjusting time estimates of selected critical activities may be all that is needed. With the revised estimates, the scheduling calculation process begins again. However, negotiating revised time estimates may meet with reluctance. If a stalemate prevents achieving acceptable or credible estimates, this may be the time to employ the three-time estimate approach.

Once the project is under way, a periodic status review may reveal several activities that are behind schedule. To keep the project on target, there will be an effort to adjust the timing of the most critical activities. Depending on the severity of the timing problem, all critical activities may be analyzed and adjusted. The search for a solution may also extend to revising the project plan.

In other situations, when critical jobs are finished ahead of schedule, check on rescheduling some future critical activities that appear in danger of not being completed on time. A popular alternative may be to leave the estimates of the succeeding critical activities intact and forecast an earlier completion of the project.

ESTIMATING RULES

Maintaining consistency requires application of specific rules when identifying the time estimates on the project planning diagram:

- Place time estimates (the activity time durations) on the bottom side of the arrow.
- Use the same unit of time for the entire project.
- Show the unit of time with each time estimate or in the legend of the diagram.
- Use whole numbers. (Whole numbers may not be required in computerized scheduling. Most computer programs will accept decimals, such as parts of a day, week, or month.)

MANUAL TIMING CALCULATIONS

Time estimates, the basis for calculating the scheduled timing, need to be placed under each activity on the project planning diagram before beginning the schedule calculations. The schedule for each project activity will initially consist of its earliest start and latest finish times and total float value. Once the earliest start computations are complete, the project duration can be evaluated.

Calculating the Earliest Start Time

The **earliest start time** is the earliest possible time an activity can begin without interfering with the completion of any of the preceding activities.

Use the following guides when calculating the earliest start times for the project activities:

- Commence calculating earliest start times with the beginning node of the project planning diagram (Figure 3.1). The scheduled time of the beginning node is set at 0 weeks. (Since all of the scheduled times are in weeks, this unit of time will not be shown in our calculations but noted in the legend of the diagram.)

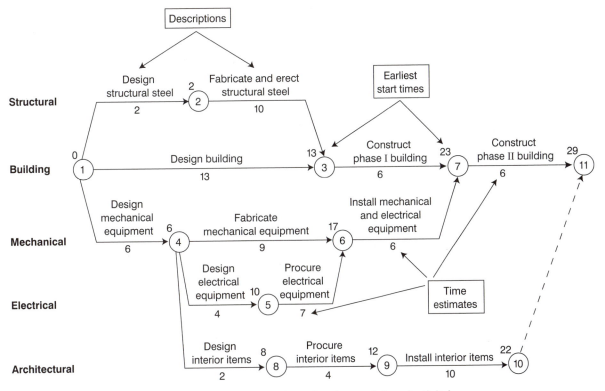

FIGURE 3.1 Time estimates and earliest start times for the work item (activity) descriptions for the Christopher Design/Build Project

- If only one arrow leads into a node, earliest start time for activities starting at that node is determined by adding the earliest start time of the preceding activity to the time estimate for the same preceding activity. *Procure interior items* starting at Node 8 is **8**. It is the sum of the earliest start time (6) for *Design interior items* that starts at node 4, and the time estimate (2) for *Design interior items*.

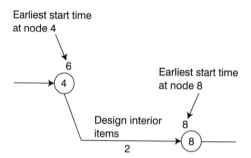

- If more than one arrow leads into a node, the earliest start time calculation is made through each of the activities as noted in Figure 3.1 and the following drawing.

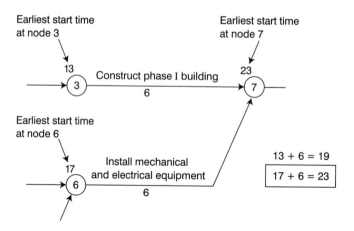

- The largest sum total is the earliest start time for the node. In this example, the path through *Install mechanical and electrical equipment*, the larger of the two paths, totals 23 at Node 7, and becomes the earliest start time of *Construct phase II building*.

Using this approach, all of the earliest start times for activities in the Christopher Design/Build Project are calculated and added in Figure 3.1.

Calculating the Project Duration

Adding the activities on the longest path (the critical path) produces a total of 29 weeks. This is the project duration for the Christopher Design/Build Project based upon the estimates used. This value of 29 weeks, the earliest start of the end node 11, is also defined as the duration of this project. This is the first chance to determine whether the plan and schedule developed so far meet the objectives for project duration established earlier in the program. It suggests that a viable plan and schedule have been assembled—management had chosen a start date of October 2, 1995 and wanted completion by April 23, 1996—29 weeks.

Calculating the Latest Finish Time

The latest finish time is the latest time an activity can be completed without delaying the end of the project. These guides should be applied in determining the latest finish times for a project:

*   The project duration must first be determined by calculating the early start times.
*   The project duration is the latest finish time (as well as the earliest start time) of the end node of the project.
*   The calculation of latest finish times involves working from the end node back through each node to the first node in the project.

If the duration for a project is 29 weeks and the last activity (7,11—*Construct phase II building*) requires 6 weeks, the latest finish time for activities coming into node 7 is 23.

If more than one arrow originates at a node, the latest finish time is calculated via each arrow and the smallest result is through each node to the end of the diagram.

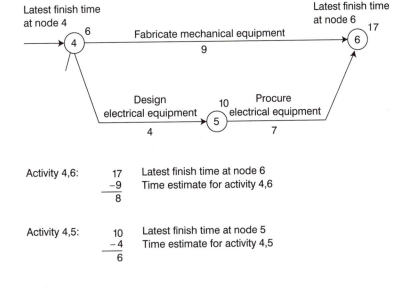

Activity 4,6:	17	Latest finish time at node 6
	−9	Time estimate for activity 4,6
	8	

Activity 4,5:	10	Latest finish time at node 5
	−4	Time estimate for activity 4,5
	6	

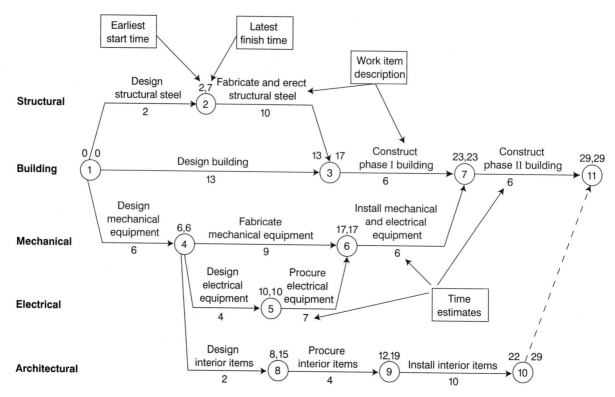

FIGURE 3.2   Time estimate, earliest start, and latest finish times for the activities of the Christopher Design/Build Project

The latest finish time for the activities that lead into node 4 is six weeks, the smaller of the two results. Were the latest finish time for activities coming into node 4 set at eight weeks, there would not be enough time remaining to complete activity 4,5 by its required latest finish time of 10.

Figure 3.2 notes the latest finish times of the project activities. The plan and schedule thus far indicate that timing objectives can be met, so we can continue without any changes or adjustments. If timing objectives were not met, we would initially examine the critical items to determine if possible adjustments would resolve the timing problem. If the project duration is still in jeopardy, then timing adjustments may have to extend to additional items in the project plan. This could require a review of the optional start and finish times for all of the project items.

Float: Optional Start and Finish Times

The float feature is one of the most important facets in project management. Its importance in project scheduling is that it identifies the activities that have optional starting and finishing dates. These activities have total float, which is

FIGURE 3.3
Calculating total float

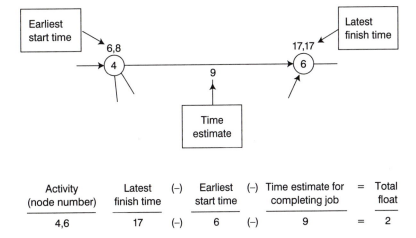

Activity (node number)	Latest finish time	(–)	Earliest start time	(–)	Time estimate for completing job	=	Total float
4,6	17	(–)	6	(–)	9	=	2

the difference between the time available for performing a job and the time required for doing it.

available time = latest finish time – earliest start time

required time = time estimate for completing the activity

Another way to express the total float is shown in Figure 3.3. The time available for Activity 4,6 (*Fabricate mechanical equipment*) is the difference between the latest finish time and the earliest start time. The time required (time estimate) to perform Activity 4,6 is subtracted from the time available to determine the total float.

Total float is a most important part of the scheduling process, especially in controlling the project. Floats are examined very closely to determine the effect of critical items when the project shows signs of "drifting." Any project activities that have 0 float (no available float time) must stay on the prescribed schedule—any delay of those activities directly affects the project duration date. On the other hand, activities that have float time available will have start and finish dates within a time range amounting to their floats. The total float figures for the activities of the Christopher Design/Build Project are shown in Figure 3.4.

In summary: The total float of each activity uses the earliest start and latest finish dates and the duration to determine its value. An activity with no float available is identified as critical, meaning no schedule flexibility. Activities with float time will have optional starting and finishing times.

TABULATING THE SCHEDULE

For a complete schedule each project activity needs, in addition to the earliest start and latest finish dates, the latest start and earliest finish dates. All of these

Activity (Node Numbers)	Description	Latest Finish Time	(−)	Earliest Start Time	(−)	Time Estimate for Completing Activity	(=)	Total Float
1-4	Design mechanical equipment	6		0		6		0*
4-5	Design electrical equipment	10		6		4		0*
5-6	Procure electrical equipment	17		10		7		0*
6-7	Install mechanical and electrical equipment	23		17		6		0*
7-11	Construct phase II building	29		23		6		0*
1-2	Design structural steel	7		0		2		5
1-3	Design building	17		0		13		4
2-3	Fabricate and erect structural steel	17		2		10		5
3-7	Construct phase I building	23		13		6		4
4-6	Procure electrical equipment	17		6		9		2
4-8	Design interior items	15		6		2		7
8-9	Procure interior items	19		8		4		7
9-10	Install interior items	29		12		10		7

*Critical path items.

FIGURE 3.4 Total float tabulation for the Christopher Design/Build Project

dates will provide all of the available scheduling options. Using the float values of each activity, the following formulas can calculate the optional starting and finishing dates:

latest start = earliest start + total float

earliest finish = latest finish – total float

Upon completing the above calculations for each activity, the total schedule for the Christopher Design/Build Project is tabulated (see Figure 3.5).

SETTING UP THE CALENDAR SCHEDULE

The tabulated schedule can be converted into a calendar schedule, which will make it more presentable for reporting and more convenient in monitoring the project. This type of schedule will be helpful in preparing the bar chart used to graphically present the schedule.

The tabulated schedule dates can be translated into calendar dates using a calendar showing the months and years of the project. Start by setting the time of the first node, 0 weeks, with the selected start date of the project. For this Christopher Design/Build Project the start date is 10/2/95, which is also the earliest start date of *Design structural steel*. Continuing to use the calendar, the earliest finish date of *Design structural steel* (time two weeks) is 10/16/95. (See Figure 3.6 for the Christopher Design/Build Project calendar schedule.)

Another helpful device is a date calculator, one version of which is shown in Figure 3.7. Set the arrow at the designated starting date for time 0, which is October 2 for this project. Keep the arrow in this location to determine the remaining dates for this project. The inner circle shows elapsed time in weekly units, while the outer circle shows daily units. By using the tabulated schedule for the remainder of the early and late times, you can complete the calendar schedule. (Office supply stores usually carry these date calculators.)

CONSTRUCTING THE BAR CHART TIME SCHEDULE

While the project planning diagram is the standard for planning the project, the bar chart is more acceptable and definitely more readable for showing the project schedule. A bar chart time schedule has these decided advantages:

- Projects are displayed effectively.
- Project activities behind schedule are readily noticed on a bar chart.
- Completion dates are specifically noted.
- Critical jobs are displayed and highlighted.
- Noncritical jobs with their float values are easily shown.

i-j	Description	Time (Weeks)	Earliest Start	Earliest Finish	Latest Start	Latest Finish	Total Float
1-2	Design structural steel	2	0	2	5	7	5
1-3	Design building	13	0	13	4	17	4
1-4	Design mechanical equipment	6	0	6	0	4	0*
2-3	Fabricate and erect structural steel	10	2	12	7	17	5
3-7	Construct phase I building	6	13	19	17	23	4
4-5	Design electrical equipment	4	6	10	6	10	0*
4-6	Fabricate mechanical equipment	9	6	15	8	17	2
4-8	Design interior items	2	6	8	13	15	7
5-6	Procure electrical equipment	7	10	17	10	17	0*
6-7	Install mechanical and electrical equipment	6	17	23	17	23	0*
7-11	Construct phase II building	6	23	29	23	29	0*
8-9	Procure interior items	4	8	12	15	19	7
9-10	Install interior items	10	12	22	19	29	7
10-11	Dummy	0	22	22	29	29	7

*Critical path items.

FIGURE 3.5 Tabulated schedule for the Christopher Design/Build Project

i-j	Activity Description	Time (Weeks)	Earliest Start	Earliest Finish	Latest Start	Latest Finish	Float (Weeks)
1-2	Design structural steel	2	10/2/95	10/16/95	11/6/95	11/20/95	5
1-3	Design building	13	10/2/95	11/1/96	10/30/95	1/29/96	4
1-4	Design mechanical equipment	6	10/2/95	11/13/95	10/2/95	11/13/95	0*
2-3	Fabricate and erect structural steel	10	10/16/95	12/25/95	11/20/95	1/29/96	5
3-7	Construct phase I building	6	1/1/96	2/12/96	1/29/96	3/11/96	4
4-5	Design electrical equipment	4	11/13/95	12/11/95	11/13/95	12/11/95	0*
4-6	Fabricate mechanical equipment	9	11/13/95	1/15/96	11/27/95	1/29/96	2
4-8	Design interior items	2	11/13/95	11/27/95	1/1/96	1/15/96	7
5-6	Procure electrical equipment	7	12/11/95	1/29/96	12/11/95	1/29/96	0*
6-7	Install mechanical and electrical equipment	6	1/29/96	3/11/96	1/29/96	3/11/96	0*
7-11	Construct phase II building	6	3/11/96	4/22/96	3/11/96	4/22/96	0*
8-9	Procure interior items	4	11/27/95	12/25/95	1/15/96	2/12/96	7
9-10	Install interior items	10	12/25/95	3/4/96	2/12/96	4/22/96	7

*Critical path items.

FIGURE 3.6   Calendar schedule for the Christopher Design/Build Project

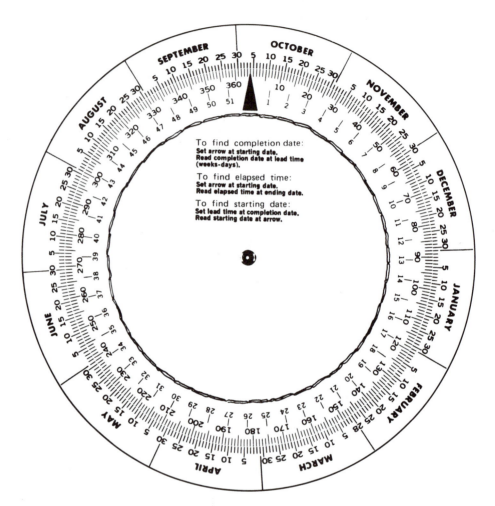

To find completion date:
**Set arrow at starting date.
Read completion date at lead time
(weeks-days).**

To find elapsed time:
**Set arrow at starting date.
Read elapsed time at ending date.**

To find starting date:
**Set lead time at completion date.
Read starting date at arrow.**

FIGURE 3.7   Date calculator (designed and produced by PERRYGRAF, Northridge, CA 91324-3552)

The bar chart time schedule can be plotted once the tabulated schedule is completed. The procedure for constructing this bar chart (which is the basic bar chart using the early start/early finish schedule) is as follows:

- Set up a date format for the project.
- Use the earliest start time for each projected activity.
- The length of each bar is the duration of each activity.
- Plot one activity per line. (In some cases it may be advantageous to plot a number of activities that are along the same path in the project planning diagram on the one line. Another group to plot on the same line may be the critical path. If the critical path items are shown on

one line, a change in the planned schedule of each item and the effect on the project duration can be readily noted.)

In the controlling phase (discussed in the next chapter), bar charts used to analyze the project with contractors or suppliers should show only the early start/early finish schedule. Showing float times, late start, and late finish schedules may give them a false sense of security that they have adequate time to finish the job, which may not be the case. On the other hand, presenting a bar chart time schedule with float times on the project activities to the owners, management, and others directly involved will present a complete picture of the project. The float times (with latest finish dates) on the chart will help give the viewer a good understanding of the project schedule. Figure 3.8 is representative of a bar chart for the Christopher Design/Build Project.

THE THREE-TIME ESTIMATE APPROACH

Individuals supplying time estimates may inject a bias in their estimates based on unfortunate experiences on past similar projects. If they expect uncertainties on this project, that unforeseen circumstances may adversely affect it, they do not want to make the same forecasting mistakes. Their estimates may be higher than acceptable, however. To offset a potential timing distortion, the **three-time estimate method** tries to mitigate this bias. Its formula requires three time estimates—optimistic, normal, and pessimistic—for each activity to calculate the expected time.

$$\text{expected time} = \frac{(\text{optimistic time}) + 4(\text{normal time}) + (\text{pessimistic time})}{6}$$

Optimistic time is the shortest possible time required for completing an activity. It means that everything goes as planned: very few design changes are made, deliveries are on time, machines and equipment operate with minimal breakdowns, personnel work within the standards, absences are infrequent, weather conditions are ideal. **Normal time** is the average schedule for this particular activity performed a number of times under similar conditions. (Normal time is usually the value used for the project activity when using just one estimate, with little or no contingency.) **Pessimistic time** is the maximum possible time required to complete an activity. An exceptional number of design changes, delivery difficulties, work delays, excessive equipment breakdowns, accidents, and miserable weather conditions are among factors that increase the time required.

The three-estimate method does have a weakness: there is a tendency to be overly pessimistic, causing the expected time to be biased toward the pessimistic time. To avoid this, it may be necessary to review more thoroughly the pessimistic times, possibly documenting reasons with the person supplying the

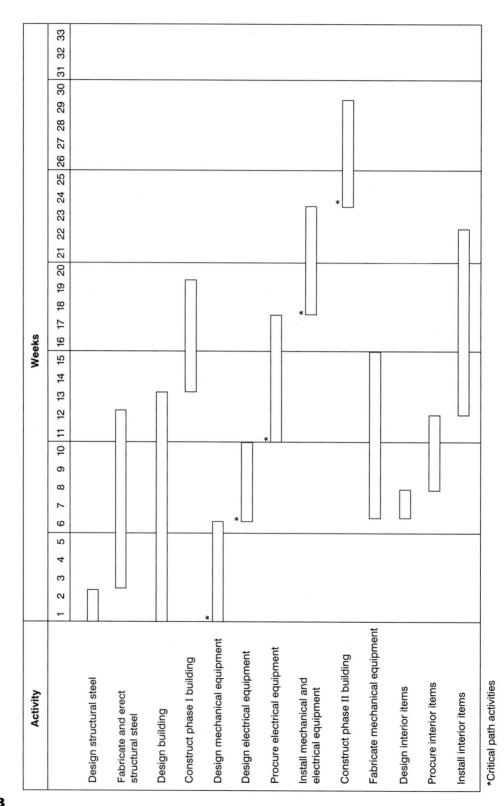

*Critical path activities

FIGURE 3.8 Bar chart time schedule for the Christopher Design/Build Project

estimate. Use the **three-time estimate method** only when the uncertainties of the project make estimates such a problem that using just one time estimate would produce unrealistic results.

Let us assume that our Christopher Design/Build Project is such a situation. Developing the three time estimates begins by using the same project planning diagram and the same time estimates that produced the initial schedule. The next step is to develop the optimistic, scheduled (also termed normal or most likely), and pessimistic times for the critical path items in the project. The critical path is used because the critical path items make up the longest path in the project— adding them up produces the duration of the project. (Some projects may have more than one critical path; in these, it will be necessary to check each path.)

For the Christopher Design/Build Project, the project team produced the optimistic, normal, and pessimistic time estimates (see Figure 3.9). An example of the time estimates developed for the activity *Design mechanical equipment* would be:

optimistic time: 5 weeks (estimated)
pessimistic time: 9 weeks (estimated)
scheduled time: 6 weeks (normal—used on original schedule)

$$\text{expected time} = \frac{(5) + (4 \times 6) + (9)}{6} = 6.3 \text{ weeks}$$

The following table summarizes the development of the expected times (based on the above formula) for each of the critical activities and for the project:

	Time (Weeks)			
Activity	Optimistic	Most Likely	Pessimistic	Expected
Design mechanical equipment	5	6	9	6.3
Design electrical equipment	2	4	8	4.3
Procure electrical equipment	4	7	9	6.8
Install mechanical and electrical equipment	5	6	7	6.0
Construct phase II building	5	6	7	6.0
		29		29.4

From the preceding table, we see that the sum of the critical items of the two critical paths equals their respective durations:

scheduled (or most likely) time = 29 weeks

expected time = 29.4 weeks

The project completion, based upon the three time estimates used for each activity, is forecasted to be 0.4 weeks or three days later than the scheduled date of April 22, 1996. This would suggest that the original estimates were plausible.

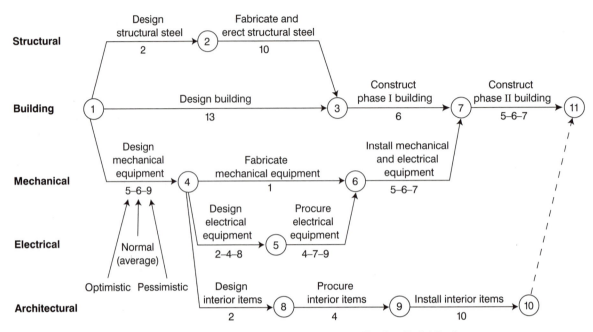

FIGURE 3.9   Three time estimates applied to the Christopher Design/Build Project

PROBABILITY OF MEETING SCHEDULED DATES

Determining the probability of meeting a schedule is helpful for resolving problems caused by uncertainty. It is only a tool, however, and the real issue is determining what problems make a schedule less probable of being met.

With the three-time estimate method, statisticians have provided formulas for calculating the probability of meeting the project duration or even meeting the scheduled finish date of any specific activity.

The basic factors used in the three-time estimate method can be translated into a distribution curve (Figure 3.10). The example used is the activity *Design mechanical equipment*, which shows the location of the time estimates: *a* (optimistic time), *c* (pessimistic), and *b* (most likely time). It is assumed that the curve has only one peak, *b*, the most likely time for completion. The *b* represents the completion with the highest probability of occurring, while the low points, *a* and *c*, indicate dates that have a small chance of being realized.

The next step in determining probability is to calculate the variance of *Design mechanical equipment* by following these steps: **Variance** is a term that describes the uncertainty associated with how much time an activity will require to complete. If the variances are large (which indicates that the optimistic and pessimistic estimates are far apart), uncertainty about when this project activity will be completed may be significantly greater. If many of the activities have large variances, there is good reason to believe that the project duration as scheduled is in jeopardy. On the other hand, a small variance indicates relatively little uncertainty.

The **standard deviation** (δ) is a measure of the spread of a distribution. While it is defined for statistics students in more technical terms, for our

FIGURE 3.10 Time estimate distribution curve

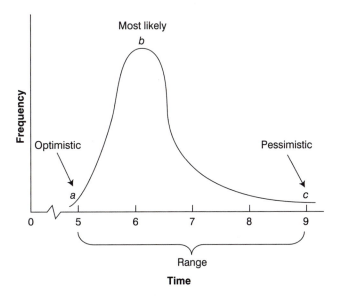

purposes, the standard deviation can be approximated as being equal to one-sixth of the range.

The range for *Design mechanical equipment:*

$$\text{Range: } 9 - 5 = 4$$

$$\text{Standard deviation: } \frac{4}{6} = 0.67$$

In statistical terms, the variance equals the standard deviation squared.

$$\text{variance} = \sigma^2$$

$$\sigma^2 = (\text{standard deviation})^2 = 0.44$$

Since we are interested in determining the uncertainty of the total project, we need to calculate the variances of project items that will affect the total project duration—the critical path items. Development of the variances of these items is shown in the following table:

Activity (Critical)	Range	Standard Deviation	Variance
Design mechanical equipment	4	0.67	0.44
Design electrical equipment	6	1.00	1.0
Procure electrical equipment	5	0.83	0.69
Install mechanical and electrical equipment	2	0.33	0.11
Construct phase II building	2	0.33	0.11

$$\text{Project variance: } \Sigma\sigma^2 = 2.36$$

$$\sqrt{2.36} = 1.53$$

The project standard deviation is calculated from the project variance, which represents variances of the critical path activities. A typical statistical table, shown in Figure 3.11, uses the project standard deviation to provide the probability information. This table, derived from the formula noted below, can determine how likely the scheduled completion date will be met.

$$Z = \frac{t_s - t_e}{\sigma} = \frac{29 - 29.4}{1.53} = \frac{-0.4}{1.53} = -0.26$$

where Z = measure related to the probability of meeting the scheduled data

   t_s = scheduled time for the activity

   t_e = expected time for the activity

   σ = standard deviation—the square root of the sum of the variances
       of the activities (critical activities)

Using the probability table: When the value of Z = -0.26, the probability is 0.397. This value indicates that there is a 39.7% chance of meeting the scheduled date of 29 weeks for the project.

A statistical exercise of this type helps the project team realize that this project will need special attention to be completed on time. The team will need to review certain critical path items, especially those with significantly biased pessimistic times, then revise these, if possible, and recalculate the expected times of the affected items. The objective is to keep the project on schedule, and additional iterations should be done until the project is back on schedule.

This exercise can be done not only during the scheduling procedure prior to start of the project, but during the course of the project when further uncertainties arise.

SCHEDULING WITH THE COMPUTER

Once you understand the scheduling terminology and analysis explained in this chapter, and the project planning approach explained in the previous chapter, you can use project management software to schedule a project.

The planning diagram (Figure 3.1) for the sample Christopher Design/Build Project contains the required computer input data—activity description with node numbers, milestones, and activity durations—needed to generate the schedule reports. Using a worksheet facilitates the entering of computer input data, making it most useful and efficient (and advisable). The worksheet is fashioned after the appearance of the computer input screen when it is set up to accept data from the project planning diagram. After the data is placed on the worksheet, if any errors are detected the corrections before data entry will be a time-saver. The **activities/resources** worksheet includes the activities timing and milestones input and the resources input screens. (Resource

FIGURE 3.11
Probability table

Z	Probability	Z	Probability
0.0	0.5000	−3	0.0013
0.1	0.5398	−2.9	0.0019
0.2	0.5793	−2.8	0.0026
0.3	0.6179	−2.7	0.0035
0.4	0.6554	−2.6	0.0047
0.5	0.6915	−2.5	0.0062
0.6	0.7257	−2.4	0.0082
0.7	0.7580	−2.3	0.0107
0.8	0.7881	−2.2	0.0139
0.9	0.8159	−2.1	0.0179
1.0	0.8413	−2.0	0.0228
1.1	0.8643	−1.9	0.0287
1.2	0.8849	−1.8	0.0359
1.3	0.9032	−1.7	0.0446
1.4	0.9192	−1.6	0.0548
1.5	0.9332	−1.5	0.0668
1.6	0.9452	−1.4	0.0808
1.7	0.9554	−1.3	0.0968
1.8	0.9641	−1.2	0.1151
1.9	0.9713	−1.1	0.1357
2.0	0.9772	−1.0	0.1587
2.1	0.9821	−0.9	0.1841
2.2	0.9861	−0.8	0.2119
2.3	0.9893	−0.7	0.2420
2.4	0.9918	−0.6	0.2743
2.5	0.9938	−0.5	0.3085
2.6	0.9953	−0.4	0.3446
2.7	0.9965	−0.3	0.3821
2.8	0.9974	−0.2	0.4207
2.9	0.9981	−0.1	0.4602
3.0	0.9987	−0.0	

abbreviations only are shown at this time with resource levels added after discussions in Chapter 6.) It will not be necessary to show the numbered nodes for the activities and milestones in any consecutive order. When completed the activities/resources worksheet for the sample project will appear as shown in Figure 3.12.

To explain the use of the computer for scheduling we will use the IN CONTROL! software included with this book. (Appendix A contains a more complete IN CONTROL! User's Guide; however, we are summarizing selected sections to maintain continuity in explaining the computer application.)

Activities/Resources Worksheet

Project Name:
Christopher Design/Build Project

Start Node	End Node	Duration (Days)	Description	Resources (Daily Requirements)			
				EN	AR	DE	SP
Start	1	—	———————	—	—	—	—
1	2	14	Design structural steel	2	—	6	—
1	3	91	Design building	6	4	12	—
1	4	42	Design mechanical equipment	4	—	4	—
2	3	70	Fabricate and erect structural steel	1	—	2	10
3	7	42	Construct phase I building	3	—	—	20
4	5	28	Design electrical equipment	4	—	8	—
4	6	63	Fabricate mechanical equipment	2	—	4	—
4	8	14	Design interior items	—	4	8	—
5	6	49	Procure electrical equipment	2	—	1	—
6	7	42	Install mechanical and electrical equipment	4	—	—	16
7	11	42	Construct phase II building	5	—	—	20
8	9	28	Procure interior items	—	2	2	—
9	10	70	Install interior items	—	2	—	12
10	11	—	Dummy	—	—	—	—
11	F	—	———————				
1	Milestone		Start project				
3	Milestone		Complete structural steel				
6	Milestone		Complete electrical equipment–	EN—Engineers			
			design and procure	AR—Architects			
7	Milestone		Complete mechanical and electrical;	DE—Designers			
			Complete phase I	SP—Skilled Personnel/Labor			
10	Milestone		Complete architectural				
11	Milestone		Complete project				

FIGURE 3.12 Sample worksheet for the Christopher Design/Build Project

54

1. Installing IN CONTROL!
   a. Insert the IN CONTROL! Plan disk in floppy disk drive.
   b. At C:<\\> create a subdirectory NCONTROL.
      - Type **MD NCONTROL** [ENTER]
      - Type **CD NCONTROL** [ENTER]
   c. Type **COPY A:*.*** [ENTER]
   d. When files are copied and you wish to add color, stay in the NCONTROL subdirectory.
      - At C:\NCONTROL type **AT C** [ENTER]
   e. To start the program, type **IC** [ENTER]

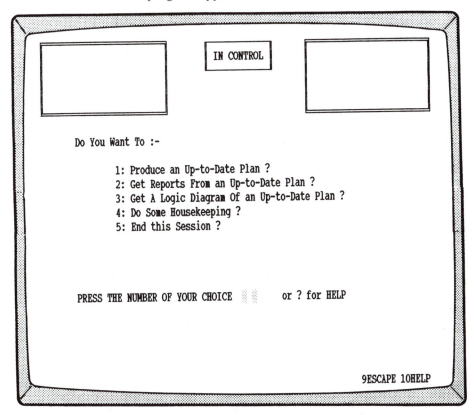

This program is generally configured and accepted by most computers. If you have questions, refer to the configuration section in Appendix A.

Using the Computer Program

Before entering into the computer, certain preliminary actions, called **house-keeping** items, are required: preparing a calendar and preparing abbreviations.

To prepare a **calendar**:

1. Press Housekeeping Option **1**: Create a New Calendar File?
2. Enter calendar name **DBPC** [ENTER].

3. Enter calendar range of project (for this project, use OCT/95 as the start and MAY/96 as the complete dates).
4. Enter the daily schedule for this project (since our sample project is on an accelerated seven-day-a-week schedule, you will include Saturday and Sunday as working days).
5. Review this calendar. If okay, then save and return to the housekeeping menu.

To prepare the abbreviations file:

1. Press Housekeeping Option **2**: Work with Abbreviation Dictionaries?
2. Press Option **1**: Create a New Abbreviations File?
3. Type **DPBA** [ENTER].
4. Enter several abbreviations:
   a.   EN—ENGINEERS
   b.   AR—ARCHITECTS
   c.   DE—DESIGNERS
   d.   SP—SKILLED PERSONNEL
5. Review the abbreviations file. If okay, save and return to the main menu.

You are now ready to input the data from the activities/resources worksheet:

1. From the main menu, choose Option **1**: Produce an Up-to-Date Plan?
2. Press **1**: A New Project Plan.

```
Please Enter The Following Information :-
----------------------------------------

 Plan File Identification (1 to 8  Letters):

 Left  Hand Report Title  (Upto 30 Letters):

 Centre    Report Title   (Upto 30 Letters):

 Right Hand Report Title  (Upto 30 Letters):

 Calendar Code-Name        (1 to 8  Letters):

 Abbreviations Code-Name  (1 to 8  Letters):

 Plan Format Type    (A=Arrow, P=Precedence):

 Time Units per Day (Numeric,Upto 3 digits):  1


 NOTE : Press RETURN only to Skip Item, or to Leave it Unchanged
        Press ESCAPE to Back Up One Step, or ? for HELP


                                          9ESCAPE 10HELP
```

3. Enter required information that will recognize the plan and print out the titles to appear on reports. ([ENTER] after each line.)
   a. **Plan File Identification**: Place name to identify plan. (For Christopher Design/Build Project, type **DBP**.)
   b. **Report Titles**: Enter descriptive titles that you wish to appear on reports. (For Christopher Design/Build Project, type **Project Management** for the left report title; **Design/Build Project** for the center, and **Boca Raton, Florida** for the right report title.)
   c. **Calendar Code-Name** and **Abbreviations Code-Name**: Type **DBPC** and **DBPA** for calendar and abbreviations file for the Christopher Design/Build Project.
   d. **Plan Format Type**: For Arrow Plan, type **A**.
   e. **Time Units per Day**: Type **1** for single-day unit. Respond **Y** to the prompt *All Items Correct (Y or N)* if satisfied with entries of the identification data.

ENTERING THE PROJECT PLAN ACTIVITIES

Once Y (Yes) is typed, a screen appears that accepts the project activity data noted on the project planning diagram or, as recommended, from the activities/resources worksheet. This is the completed screen after it has accepted all of the data.

```
COMMAND?                          Current Plan: DBP      version 5

 LINE    FROM    TO    DURATION   DESCRIPTION                          ?=HELP

  1 :   START-> 1                 NOT BEFORE DAY     2
  2 :     1  -> 2        14       DESIGN STRUCTURAL STEEL
  3 :        -> 3        91       DESIGN BUILDING
  4 :        -> 4        42       DESIGN MECHANICAL EQUIPMENT
  5 :     2  -> 3        70       FABRICATE & ERECT STRUCTURAL STEEL
  6 :     3  -> 7        42       CONSTRUCT PHASE I BUILDING
  7 :     4  -> 5        28       DESIGN ELECTRICAL EQUIPMENT
  8 :        -> 6        63       FABRICATE MECHANICAL EQUIPMENT
  9 :        -> 8        14       DESIGN INTERIOR ITEMS
 10:     5  -> 6        49       PROCURE ELECTRICAL EQUIPMENT
 11:     6  -> 7        42       INSTALL MECH & ELECT EQUIP
 12:     7  -> 11       42       CONSTRUCT PHASE II BUILDING
 13:     8  -> 9        28       PROCURE INTERIOR ITEMS
 14:     9  -> 10       70       INSTALL INTERIOR ITEMS
 15:    10  -> 11        0       DUMMY
 16:    11  ->FINISH            UNSCHEDULED


   1DESCRP 2RESRCE 3INSERT 4NEW PG 5HDINGS 6GO TOP 7ZOOM  8BARCHT 9ESCAPE 10HELP
```

To review the input:

1. Each line number accepts one activity of the project. Line 1 is common to all projects; it denotes start of the project.
2. The start and finish node numbers will be in the From and To columns, respectively.
   a.   Under From, Enter **S** [ENTER].
   b.   Under To, Type **1** (the first Node number).
   c.   Enter **Y**; enter **2** (project start is scheduled on the 1st working day of the project).
3. Line 2 accepts the project data for the first activity.
   a.   Type the start node number [ENTER].
   b.   Type the finish node number [ENTER].
   c.   Type number of days in the Duration column [ENTER].
   d.   Type **Description** [ENTER].
4. Follow the same procedure to enter all of the activities.
5. Line 16 denotes the finish data for this project. (Finish data can be done on any line; however, it must be for the last node number of the last activity of the project.)
   a.   Under From, type **11** [ENTER].
   b.   Under To, type **F** (Finish will be displayed).
   c.   Type **N** to indicate that the finish date is yet to be scheduled.
6. When all of the data is entered, type [ENTER], then press [ESC].
   a.   Type **2** [ENTER] at *Enter a New Time Now Day Number*. (This corresponds to the starting date of the project.)
   b.   Press [ESC] to store project data and return to the main menu.

You are now ready to get schedule reports.

The Computer Schedule Report

1. From the main menu, press **2**: Get Reports from an Up-To-Date Plan?
2. Select the plan (or project) for which you want a schedule report (see screen on p. 59).
3. Press **1** for the Standard Report Listing for the tabulated schedule showing earliest and latest start and finish dates, float, and duration of each project activity.
4. Press **1**: The Earliest Activities First to show the activities in a sequence from the earliest starting date to the latest starting date.
5. Press [ENTER] until you return to the reporting options menu. Press **8**: Start or Cancel the Run, then [ENTER] to start the printer which will print a page like the one shown in Figure 3.13.
6. If you wish to preview the schedule prior to printing:
   a.   Press **6**: Screen Report.
   b.   Press **F6** to check the schedule tabulation.
   c.   Press **F9** to return to print report, then press **Y**.

```
┌─────────────────────────────────────────────────────────────────────┐
│                              ┌─────────────┐                          │
│  ┌──────────────────────┐    │ IN CONTROL  │    ┌──────────────────┐  │
│  │ Current Plan :DBP     │    └─────────────┘    │ Version : 5      │  │
│  │ Status       :LOADED  │                       │ Size    : 16     │  │
│  └──────────────────────┘                        └──────────────────┘  │
│                                                                         │
│    Choose the Next Report :-                  Reports Ordered So Far:   │
│                                                                         │
│        1: Standard Report Listing                                       │
│        2: Project/Period Barcharts                                      │
│        3: List of Activities & Relationships                            │
│        4: Complete Resources Report                                     │
│        5: Histogram for a Resource                                      │
│        6: Screen Report                                                 │
│        7: Disk Report                                                   │
│        8: Start or Cancel the Run                                       │
│                                                                         │
│    PRESS THE NUMBER OF YOUR CHOICE        or ? for HELP                 │
│                                                                         │
│                                                        9ESCAPE 10HELP   │
└─────────────────────────────────────────────────────────────────────┘
```

PROJECT MANAGEMENT   DESIGN/BUILD PROJECT   BOCA RATON, FLORIDA
     REPORT TYPE :STANDARD LISTING   PRINTING SEQUENCE :Earliest Activities First
                                                          SELECTION CRITERIA :ALL
     PLAN I.D.   :DBP   VERSION 5   TIME NOW DATE   : 2/OCT/95

	ACTIVITY DESCRIPTION	EARLIEST START	EARLIEST FINISH	LATEST START	LATEST FINISH	DURATION	FLOAT
1-	4 DESIGN MECHANICAL EQUIPMENT	2/OCT/95	12/NOV/95	2/OCT/95	12/NOV/95	42	0 *
1-	3 DESIGN BUILDING	2/OCT/95	31/DEC/95	30/OCT/95	28/JAN/96	91	28
1-	2 DESIGN STRUCTURAL STEEL	2/OCT/95	15/OCT/95	6/NOV/95	19/NOV/95	14	35
2-	3 FABRICATE & ERECT STRUCTURAL STEEL	16/OCT/95	24/DEC/95	20/NOV/95	28/JAN/96	70	35
4-	5 DESIGN ELECTRICAL EQUIPMENT	13/NOV/95	10/DEC/95	13/NOV/95	10/DEC/95	28	0 *
4-	6 FABRICATE MECHANICAL EQUIPMENT	13/NOV/95	14/JAN/96	27/NOV/95	28/JAN/96	63	14
4-	8 DESIGN INTERIOR ITEMS	13/NOV/95	26/NOV/95	1/JAN/96	14/JAN/96	14	49
8-	9 PROCURE INTERIOR ITEMS	27/NOV/95	24/DEC/95	15/JAN/96	11/FEB/96	28	49
5-	6 PROCURE ELECTRICAL EQUIPMENT	11/DEC/95	28/JAN/96	11/DEC/95	28/JAN/96	49	0 *
9-	10 INSTALL INTERIOR ITEMS	25/DEC/95	3/MAR/96	12/FEB/96	21/APR/96	70	49
3-	7 CONSTRUCT PHASE I BUILDING	1/JAN/96	11/FEB/96	29/JAN/96	10/MAR/96	42	28
6-	7 INSTALL MECH & ELECT EQUIP	29/JAN/96	10/MAR/96	29/JAN/96	10/MAR/96	42	0 *
7-	11 CONSTRUCT PHASE II BUILDING	11/MAR/96	21/APR/96	11/MAR/96	21/APR/96	42	0 *

FIGURE 3.13   Project schedule report for the Christopher Design/Build Project

Bar Chart Schedules

When you have returned to the reporting options menu and you would like to see a bar chart schedule, press **2**: Project/Period Barcharts.

1. There are two types of schedules:
   a. The **Project Barchart** compresses the total project on one page.
   b. The **Period Barchart** will print the total project on a daily basis over several pages.
2. For a project bar chart that is more desirable for our immediate purposes, type **(0)** at *Type the Day Number That You Wish This Report to Start From*. For the period bar chart, if desired, type **2**, the starting date of the project.
3. Press [ENTER] until you return to reporting options menu, then press **8**: Start or Cancel the Run and [ENTER]. Your resulting chart will look like Figure 3.14.

Using the computer for analysis and making changes and/or additions is superior to the manual method of calculating the schedule. As soon as changes are made to the input screen, printed or screen reports can be instantaneously created, saving considerable amounts of time. Before entering the input data, you should still check the project duration by making an initial pass through the critical path. Manually adding the durations of the critical items on the planning diagram will produce the duration of the total project. This can be a time-saver if all of the project changes are made before the data is stored in the computer.

```
PROJECT MANAGEMENT                                    DESIGN/BUILD PROJECT                                      BOCA RATON, FLORIDA
       REPORT TYPE :COMPRESSED PERIOD BARCHART                              PRINTING SEQUENCE :Earliest Activities First
                                                                           SELECTION CRITERIA :ALL
       PLAN I.D.   :DBP      VERSION  5                                     TIME NOW DATE     : 2/OCT/95
=====================================================1995=========================1996==============================================
   PERIOD COMMENCING DATE           !2    !27   !21   !16   !10   !4    !29   !25   !19   !14   !8    !3    !
   MONTH                            !OCT  !     !NOV  !DEC  !JAN  !FEB  !     !MAR  !APR  !MAY  !JUN  !JUL  !
   PERIOD COMMENCING TIME UNIT      !2    !27   !52   !77   !102  !127  !152  !177  !202  !227  !252  !277  !
====================================================================================================================================
   1-   4 DESIGN MECHANICAL EQUIPMENT  !CCCCCC!CCCCC !     !     !     !     !     !     !     !     !     !
   1-   3 DESIGN BUILDING              !======!======!======!=====..!......!     !     !     !     !     !     !
   1-   2 DESIGN STRUCTURAL STEEL      !====..!......!.     !     !     !     !     !     !     !     !     !
   2-   3 FABRICATE & ERECT STRUCTURAL S !   ===!======!=====!===....!.....!     !     !     !     !     !     !
   ----------------------------------------------------------------------------------------------------------------------------
   4-   5 DESIGN ELECTRICAL EQUIPMENT   !     !   CC!CCCCCC!     !     !     !     !     !     !     !     !
   4-   6 FABRICATE MECHANICAL EQUIPMENT !    !   ==!======!======!==... !     !     !     !     !     !     !
   4-   8 DESIGN INTERIOR ITEMS         !     !   ==!==....!......!..    !     !     !     !     !     !     !
   8-   9 PROCURE INTERIOR ITEMS        !     !     !   ====!===....!......!....!     !     !     !     !     !
   ----------------------------------------------------------------------------------------------------------------------------
   5-   6 PROCURE ELECTRICAL EQUIPMENT  !     !     !   C!CCCCCCC!CCCCC !     !     !     !     !     !     !
   9-  10 INSTALL INTERIOR ITEMS        !     !     !     !   ====!======!======!==....!......!..  ' !     !     !
   3-   7 CONSTRUCT PHASE I BUILDING    !     !     !     !   ==!======!===...!...   !     !     !     !     !
   6-   7 INSTALL MECH & ELECT EQUIP    !     !     !     !   CC!CCCCCC!CCCC  !     !     !     !     !     !
   ----------------------------------------------------------------------------------------------------------------------------
   7-  11 CONSTRUCT PHASE II BUILDING   !     !     !     !     !     !   CCC!CCCCCCC!C  !     !     !     !
====================================================================================================================================
Barchart Key:-  CCC :Critical Activities  === :Non Critical Activities  NNN :Activity with neg float  ... :Float
```

FIGURE 3.14   Bar chart schedule for the Christopher Design/Build Project

SUMMARY

Scheduling concerns timing: how much time is required versus the time available to complete an activity. Getting time estimates for completing each activity in the project is the first step in the scheduling process. Whenever possible, an experienced person provides the time estimate under the assumption the project activity will be performed on a normal basis.

Scheduling begins when each estimate is placed under its respective project activity on the project planning diagram. Initial calculations are for the earliest start time and latest finish time of each activity. (For expediency, the earliest start times and latest finish times are placed on their appropriate nodes on the project planning diagram.)

The earliest start time of the last node also denotes the project duration and is checked against project objectives. If this initial effort indicates difficulty in meeting the project objectives, the process requires additional calculations of the earliest start times before evaluation of the project can proceed. Float values are derived from the earliest start times and latest finish times for activities that have optional starting and finishing dates, if any. Float is defined as the difference between the time required to complete an activity and the time available to complete this same activity. Activities with 0 float values are critical; when joined together, these make up the longest path (or "critical path") through the project. The earliest start time of the end node of the last item in the critical path is the duration of the project.

In addition to preparing a schedule tabulation of the project showing the earliest and latest start dates for each activity, earliest and latest finish dates, and available float, a bar chart to show these values is an excellent graphic element.

Where estimates are surrounded by uncertainty, the use of three time estimates—optimistic, normal, and pessimistic—can possibly bring uncertainties into a clearer focus. The three-time estimate approach helps address apprehension about the time estimates and produces a more realistic project completion date.

The probability of achieving a scheduled completion date of a project can be developed by using three time estimates for those items on the critical path and employing a statistical approach. This method is useful for selective projects but is not necessary for most.

For small projects, schedules can be calculated manually; relatively large projects benefit from project management software to calculate schedules. For all projects, the manual method is a good way to determine initially whether the project duration date is satisfactory before proceeding with the relatively longer task of collecting all the necessary computer input data.

Benefits from schedule calculations include:

- establishing a credible project duration
- identifying the critical items in the project
- identifying jobs that have scheduling flexibility

Exercises

1. If planning a project is determining what has to be done, then what is scheduling a project?
2. Name three benefits when scheduling a project in the manner prescribed in this chapter.
3. Using the planning diagram shown in Figure 1.2, add node numbers for the project activities and place the estimates below on the diagram:

Activity	Weeks
Structural engineering	3
Fabricate and erect structural steel	10
Construction engineering	5
Phase I construction	12
Phase II construction	20
Award contract	4
Equipment engineering	8
Fabricate and deliver equipment	12
Phase I installation	10
Phase II installation	8
Long lead items—engineering	6
Long lead items—procurement	18

    a. Calculate the earliest start and latest finish times at each node on the diagram.

    b. What is the project completion time?

    c. Identify the critical path on the project planning diagram.

4. Prepare a tabulated schedule for the Exercise 3 project. This schedule should show the earliest and latest start dates, earliest and latest finish dates, float, and duration for each project activity.
5. Using Exercise 3, start the project on January 1, 1996. With a date calculator or 1996 calendar, develop the schedule using calendar dates.
6. Using Exercise 3, construct a bar chart time schedule showing float on all of the applicable activities.
7. Prepare a computerized schedule of the project shown in Figure 1.2 using the data in Exercise 3. For input data use the estimated durations and the same node numbering as set up in Exercise 3.
8. Prepare a bar chart for the project described in Exercise 7.
9. Prepare the computerized input data for the precedence diagramming (PDM) plan of Chapter 2 Exercise 5. (The schedule for this plan will be the same as that of the arrow diagramming [ADM] plan used in the previous exercises.)

Project Control

THIS CHAPTER WILL COVER

- Monitoring a project

- Measuring project performance

- Communicating
  a. Constructing the progress schedule
  b. Preparing a status report

Controlling a project—maintaining and measuring its performance once it gets under way—involves four stages: (1) periodically reviewing the performance of each project activity on the schedule, (2) measuring and evaluating its performance against the planned project objectives, (3) taking the necessary action on critical activities that are affecting project performance, and (4) communicating overall progress to management and project team through meetings, memos, and written reports. These reports emphasize the critical activities that are falling behind schedule and what corrective action they require.

PROJECT CONTROL: MEASURING PROJECT PERFORMANCE

Project control is an early warning system to resolve problems in the earlier stages of the project and thus avoid panic situations when the project nears completion. It compares the early start/early finish schedule, based on the original plan, with the actual performance of the project. (The early warning system approach is also used for project cost control, discussed in Chapter 5.)

The four distinct stages are characterized as follows:

1. **monitor**: periodic checking of the plan and schedule
2. **assess**: identifying the critical items and problems
3. **resolve**: finding solutions to the problems
4. **communicate**: advising management, project team members, suppliers, vendors, and contractors of project status

Monitoring the Project

Monitoring project activities is a continuous checking of their performance. This effort is essential to keeping the project under control. At least once a month, the project team will formally check the total project. If there are disturbing signs that the project may be heading for difficulties, formal monitoring and reporting will be more frequent. The present status of each project activity is noted and recorded during the monitoring stage.

For the Christopher Design/Build Project, at the end of Week 12 we meet with those involved to review the schedule. The review finds the timing of several activities has deviated from the planned schedule.

Activity	Duration (Weeks)	
	Original	Revised
Design structural steel	2	4
Design building	13	16
Design electrical equipment	4	6
Procure interior items	4	8

i-j	Description	Time (Weeks)	Earliest Start	Earliest Finish	Latest Start	Latest Finish	Total Float
1-2	Design structural steel	2̸ 4	0	2	5	7	5
1-3	Design building	1̸3̸ 16	0	13	4	17	4
1-4	Design mechanical equipment	6	0	6	0	4	0
2-3	Fabricate and erect structural steel	10	2	12	7	17	5
3-7	Construct phase I building	6	13	19	17	23	4
4-5	Design electrical equipment	4̸ 6	6	10	6	10	0
4-6	Fabricate mechanical equipment	9	6	15	8	17	2
4-8	Design interior items	2	6	8	13	15	7
5-6	Procure electrical equipment	7	10	17	10	17	0
6-7	Install mechanical and electrical equipment	6	17	23	17	23	0
7-11	Construct phase II building	6	23	29	23	29	0
8-9	Procure interior items	4̸ 8	8	12	15	19	7
9-10	Install interior items	10	12	22	19	29	7
10-11	Dummy	0	22	22	29	29	7

FIGURE 4.1   Duration changes to the project schedule for the Christopher Design/Build Project

The next steps will determine whether any of these activities will significantly affect the schedule. To get a clear picture, the revised durations of these activities are shown on the project schedule (see Figure 4.1). Those are the only markings as all of the other project activities have progressed, are progressing, or are planned to progress according to the original schedule. The project team members responsible for their respective activities review and confirm changes on the planning diagram. Memos, status meetings, or any other means of communication may serve the same purpose.

Assessing the Project

The next stage, assessing the project, determines what effect the changes have on the project schedule, particularly on the project completion date. The duration changes replace the original times on the planning diagram of the Christopher Design/Build Project (Figure 4.2). With the revised durations in place, new calculations are made for the earliest start times and latest finish times of all of the project activities. The earliest start time of Node 11, the last node of the project (which is also the duration of the Christopher Design/Build Project), has changed from 29 weeks to 31 weeks. This indicates that, because of the changes, the project has been extended two weeks beyond the original schedule.

To observe the total effect, a completely new schedule is to be prepared, requiring recalculation of the total float for each activity. The new total float table

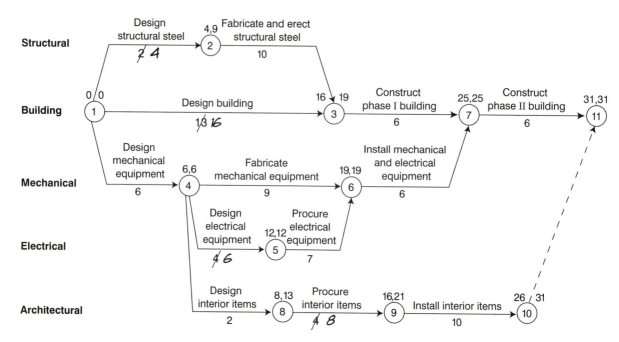

FIGURE 4.2   Revised estimates, earliest start times, and latest finish times of the Christopher Design/Build Project

will show some dramatic changes on some activities, which is a sign that a new schedule needs to be recalculated. The new schedule will be helpful to recognize all activities with 0 float and also those with relatively small float. The schedule gives more attention to these activities (1) to prevent further potential delays and (2) to reduce their durations as a possible way to restore the project to its original completion date. The revised schedule is quite complete, showing early and latest start times, early and latest finish times, durations, and float. It is very helpful in reviewing the entire project to help resolve the scheduling problems.

Resolving Problems

Certain questions must be answered to help solve the planning and scheduling concerns and/or problems of the project:

1. Which specific activities are behind schedule? How much will they delay the project date if not corrected?
2. What are the reasons for the delays that have caused the project to fall behind schedule?
3. What steps have been or are being taken (or should be taken) to get back on schedule, and what results have been achieved or are expected?
4. Will any other specific recommendations for further action restore the schedule and prevent further concerns?

The following activities noted with 0 float, especially those that contributed to schedule problems, are the first to be examined. While some of these activities were not necessarily affected by any changes, they nevertheless should be analyzed, as all of them are on the critical path that has extended the date of this project's completion to 31 weeks (2 additional weeks).

Critical Activity	Duration	Float
Design mechanical equipment	6	0
Design electrical equipment	6	0
Procure electrical equipment	7	0
Construct phase II building	6	0
Install mechanical and electrical equipment	4	0

Of the critical activities indicated, two—*Design mechanical equipment* and *Design electrical equipment*—took longer than scheduled to complete due to design changes. It is too late to address these delays, so any resolutions to correct the behind-schedule situation will need to involve other critical activities.

With two weeks needed to be eliminated from the durations of the three critical activities—*Procure electrical equipment (7), Install mechanical and electrical equipment (6),* and *Construct phase II building (6)*—these steps should be taken:

- Contact suppliers of the mechanical equipment and request deliveries one or two weeks earlier. Get firm delivery dates.
- Contact contractors of the *Phase II building* for possible work acceleration in some of their critical areas.

If these two steps are difficult to achieve, there is time to review other options. The advantage of the project control approach, using the early warning system, is that it allows time to resolve the problems—action to resolve is being taken at week 12 of the project and the duration is 17 weeks away at week 29.

Communicating

It is important to communicate the project status to management, suppliers, vendors, contractors, and project team members. While a written report is possibly one of the most valuable communication devices, other effective means include phone calls, memos, personal contact, and meeting formally as well as continuously throughout the project.

This chapter will concentrate on the importance of using a combination of charts and graphs with the written report to provide a comprehensive status report. A written periodic project status report (weekly or monthly) should be distributed to all those who want to see the project reach a successful conclusion. A status report presents a view of what is happening in the project, offering readers an opportunity to suggest improvements early enough to make an impact on the project's progress. The status report is not voluminous; it is essentially a summary. In a relatively short format the report should be readable and structured in such a way that it will likely be read. The report includes a brief background of the project, highlights critical areas, and outlines the resolutions and improvements planned.

The status report is based on the progress schedule, which is a graphic presentation of the project schedule revised to show actual progress.

CONSTRUCTING THE PROGRESS SCHEDULE

In addition to depicting schedules, a bar chart is an excellent way of tracing the progress of the project and its individual activities. Showing progress on a daily, weekly, or monthly basis can employ the same type of bar chart that was constructed for the time schedule bar chart. A chart of this type is called a progress schedule. Each individual project item with its progress can be shown separately and noted on the progress schedule.

Several rules should be followed to allow uniform interpretation of the progress schedule:

- The rectangular bars (⬜) represent the duration of the project activity. The activity farthest to the right represents the end of the project. The portion of the bar that is filled in (▬) represents how much of the activity has been completed.

- Extending the rectangular bar with a dotted bar (⬜) indicates an extension of the duration for that activity.

- Float of an activity is shown with an arrow (⬜→). The length of the arrow is the float value of the activity. (If the activity bar must be extended because of some schedule delay, it can readily be determined whether the extension will affect the project duration by its relative position to the head of the arrow.)

- The symbol for a major event date, or milestone (▽) is left hollow until the event is achieved, then it is filled in (▼).

- The planned completion date is also a milestone (▽). If the completion date needs to be extended because of time delays, the extended milestone is shown with broken lines (▽).

To illustrate a progress schedule, assume that we are reviewing the status of the Christopher Design/Build Project at the end of week 12. We secure the information from the monitoring stage, which shows the following activities have deviated from their planned schedule:

Activity	Duration (Weeks)	
	Original	Revised
Design structural steel	2	4
Design building	13	16
Design electrical equipment	7	9
Procure interior items	4	8

From the monitoring stage it was established that all of the other activities have progressed, are progressing, or are planned to progress at their planned schedule.

The early start schedule will be used for this progress schedule. Float will be shown on the progress schedule for information only. All project activities are thus expected to perform according to the early start schedule.

This type of bar chart, using the early start/early finish schedule, is distributed to vendors, suppliers, and contractors. Their activities can be controlled only if they are given no (or very little) leeway in the timing of their activities. If there are actual departures from the schedule as the project progresses, the person heading the project must decide how much float to allow. If no float is available on a specific item that is showing a potential delay, then the problem becomes more extensive. (These issues would normally have been handled in the previously discussed assess stage.) Figure 4.3 shows a progress schedule for the Christopher Design/Build Project.

PROJECT STATUS REPORT

A complete project status report will include three documents: a cover letter, executive highlights, and a summary bar chart. The cover letter contains brief statements of overall progress; executive highlights, in bullet (•) form, outline the status of the most important aspects of the project; and the summary bar chart is a graphic display of the project status. The project summary using milestones is also an effective application. (Milestones will be explained in this chapter.)

Cover Letter

The cover letter is addressed to the management as well as all other personnel with a vested interest in the project. It includes a brief description of the project, anticipated completion dates of the project and its major events (milestones),

Program report date

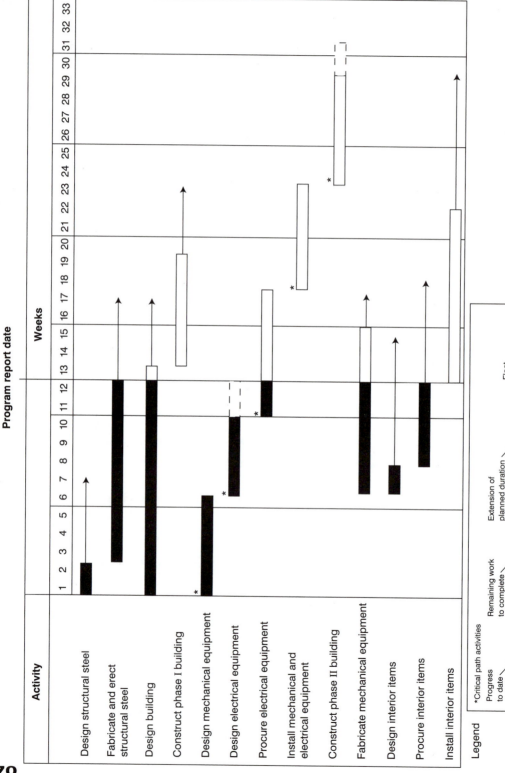

FIGURE 4.3 Progress schedule for the Christopher Design/Build Project

brief statements on the status of the project and its critical items, and potential solutions to any problems.

A cover letter avoids great detail. If additional information is needed, the reader can review the rest of the status package. Beyond that, the reader can contact the person who prepared the status report. The last line of the cover letter should say, "Please contact the writer for any comments or if additional information is needed." (See Figure 4.4 for the cover letter on progress after week 12 for the Christopher Design/Build Project.)

Executive Highlights

The second document, the executive highlights, focuses on the status of important aspects of the project. Highlights will consist of a series of bulleted or numbered statements. Each statement is a brief and specific sentence or two. Relatively small project status reports may incorporate the executive highlights in the cover letter. Figure 4.5 shows the executive highlights that would be attached to the cover letter for the Christopher Design/Build Project.

Summary Bar Chart

The summary bar chart, used for management reports, employs the late finish schedule. (The progress schedule—used for reporting progress to vendors, contractors, and suppliers—uses the early start schedule.) A summary bar chart displays all activities at their latest finish dates, allowing for no schedule flexibility. Those using this type of schedule for their work will experience potential project completion delays in the event any activity in the project "drifts." It should only be used for management and internal reporting.

The summary bar chart incorporates all the project activities, but highlights only the critical items. It avoids extraneous discussions of noncritical activities that may be delayed but have float values with enough cushion to avoid pushing back the project completion date.

The summary bar chart for the Christopher Design/Build Project is based on the work breakdown structure (WBS) shown in Figure 4.6. Management has more interest in the overview (or summary), so using the WBS fulfills this need as it is a chart showing the major divisions (or groups) of the project. The summary bar chart will use the information from the progress schedule for the current status of each activity. Activities within the same division (or group) are combined to arrive at the status of each division. The schedules of these divisions are displayed with the start at their earliest start date and the finish at their latest finish date. Activities within a division that are behind schedule can cause that division to be behind schedule, unless adequate float is available from other activities within the same division and along the same path.

Those reviewing this type of management summary bar chart will be able to perceive the progress of the total project without having to review every single

Snyder Engineering Associates
Boca Raton, Florida

December 25, 1995

Christopher Development Company
Boca Raton, Florida

Subject: December 25, 1995 Status Report
Christopher Design/Build Project

The completion of the Christopher Design/Build Project at this time is projected to be delayed two weeks from the planned occupancy of April 22, 1996 to May 6, 1996. The electrical suppliers have just informed us that there has been a delivery problem on some key component parts, and they estimate a wait of two weeks before they will get the parts.

The procurement of electrical equipment is a critical project item. While we did not appreciate receiving the above information just two days ago, we had no other alternative but to report a project completion delay at this time. However, we are optimistic that we will be able to correct this situation before our next status report at the end of January.

Our plans are to review again with the equipment suppliers urging them to honor their purchase order agreements to meet the specified dates. If there still is a problem we will meet with the personnel involved in the two subsequent critical items—*Install mechanical & electrical equipment* and *Construct phase II building*—and work on negotiating their schedules in an attempt to return the project to its original completion date.

All other project items are proceeding according to plan, with the exception of the building design and interior items delivery. While these are items with adequate float to cover the additional time they need, we are still requesting a detailed explanation on the reported delays. The program is only 40% underway and we want to discourage trends to "miss" interim objectives.

Please refer to the attached *Executive Highlights* and *Summary Progress Schedule* for further details. We welcome your comments and questions.

Sincerely yours,

Bob Sachs
Project Manager

FIGURE 4.4   Cover letter showing status of the Christopher Design/Build Project

CHRISTOPHER DESIGN/BUILD PROJECT
EXECUTIVE HIGHLIGHTS
December 25, 1995

STRUCTURAL

1. Structural steel erection proceeding on schedule. Aside from experiencing any inclement weather, all of the building steel should be completed by the end of January.
2. Phase I contractor was notified of the structural steel progress, and could start on February 1 to mobilize on the job site.

BUILDING

1. Phase I building contractor, in anticipation of starting February 1, plans to move on job site about mid-January to mobilize and to receive mechanical equipment.
2. We plan to have a joint meeting with phase I and phase II contractors about mid-January to discuss schedule improvements. We will request that they consider coordinating the construction of several critical items for the purpose of reducing the overall schedule.

MECHANICAL

1. Mechanical suppliers have all reported that their fabrication is going according to schedule.
2. Equipment shipments for the building A/C systems will begin about the third week in January when phase I contractor will be ready to receive them.

ELECTRICAL

The timing of the procurement and installation of the electrical items are critical project items. Attention to their progress is most important.

1. Supplier for electronic controls for HVAC systems reports quality problems. Several components require reordering, which is expected to delay shipments by two weeks.
2. Fabricator of in-plant boiler equipment experiencing unauthorized work stoppages. Expects settlement within a week, but will not be back to normal operations for at least one week after settlement.
3. We have assigned one person to each of the trouble areas until our equipment is completed and shipped. They will report daily on progress. They will also assist the suppliers, offering them any resources that we have available, to complete a quality product that meets our approval.

ARCHITECTURAL

1. Design of architectural interior items is complete.
2. Fabrication is essentially complete. Suppliers are requested to begin shipments about mid-January, when phase I building contractor is on the site and can accept them.

MILESTONE REPORT

Significant milestones to keep project on the 6 May 96 completion date:

1. Complete structural steel erection   12 Feb. 96
2. Complete electrical equipment procurement   12 Feb. 96
3. Complete phase I building   25 Mar. 96
4. Complete interior work   6 May 96
5. Complete phase II building; complete project   6 May 96

Snyder Engineering Associates
Boca Raton, Florida

FIGURE 4.5   Executive highlights for the Christopher Design/Build Project

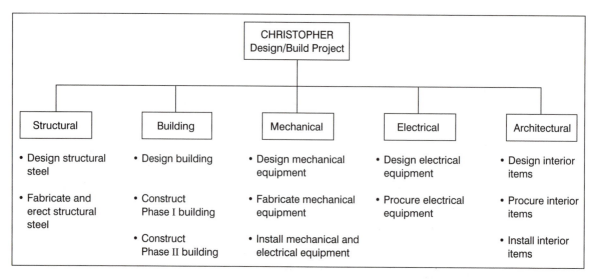

FIGURE 4.6   Work breakdown structure for the Christopher Design/Build Project

activity. The five major divisions with their planned earliest start and latest finish dates updated to Week 12 will be used in the summary bar chart. The tabulation will appear, usually with the summary bar chart, in this manner:

	Earliest Start		Latest Finish	
Major Group	Planned	Week 12	Planned	Week 12
Structural	2 Oct. 95	2 Oct. 95	28 Jan. 96	28 Jan. 96
Building	2 Oct. 95	2 Oct. 95	21 Apr. 96	5 May 96
Mechanical	2 Oct. 95	2 Oct. 95	10 Mar. 96	10 Mar. 96
Electrical	13 Nov. 95	13 Nov. 95	28 Jan. 96	28 Jan. 96
Architectural	13 Nov. 95	13 Nov. 95	21 Apr. 96	21 Apr. 96

The above schedules for the major groups are shown on the summary bar chart for the Christopher Design/Build Project (see Figure 4.7).

Because of delays in specific activities, the latest finish date for the building group is extended beyond its planned completion date. The finish date of this group has been extended from April 21, 1996 to May 5, 1996.

An option to putting the executive highlights and summary bar chart on separate pages is a comprehensive one-page report incorporating both of these documents. The executive highlights commentary appears below the summary bar chart. This type of reporting is popular for relatively small projects. Attached to the cover letter, this page may suffice as a completed project status report for distribution to management.

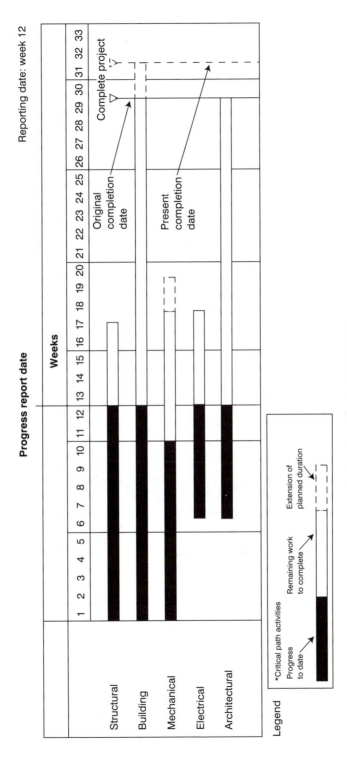

FIGURE 4.7 Summary bar chart for the Christopher Design/Build Project

MILESTONES

Another means to communicate the status of a project, especially to management, is the use of milestones. Milestones are selected events that are crucial to achieving objectives. They are key events, such as the completion date of a major phase of the project, the delivery date of a major equipment item, or the date of a key management decision. These events may or may not be on the critical path.

The milestone approach is another excellent option for summarizing project status to higher management, as it capsulizes the status of major events. The milestone report is another way to isolate behind-schedule activities so management can quickly note these problem areas and can assist in formulating corrective action early in the project.

Milestones noted on the Christopher Design/Build Project planning diagram (Figure 4.8) represent key starting and/or completion dates of the project.

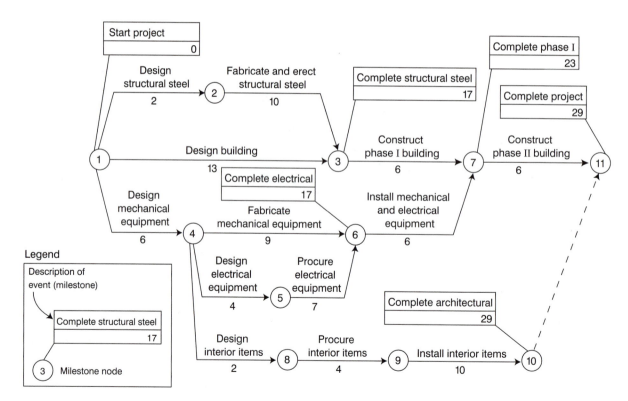

FIGURE 4.8   Milestones for the Christopher Design/Build Project

To illustrate the use of milestones for reporting status of the Christopher Design/Build Project:

1. List the major events.
2. Tabulate their original scheduled milestone dates as noted on the project planning diagram.
3. Tabulate their week 12 dates.

Milestone Schedule

Node Number	Event	Original Date	Float	Status (25 Dec. 95) Date	Float
1	Start project	2 Oct. 95	—	2 Oct. 95	—
3	Complete structural steel	31 Jan. 96	4	12 Feb. 96	3
6	Complete electrical	31 Jan. 96	0	12 Feb.96	0
7	Complete mechanical and electrical; complete phase I building	11 Mar. 96	0	25 Mar. 96	0
10	Complete architectural	22 Apr. 96	7	6 May 96	7
11	Complete project	22 Apr. 96	—	6 May 96	—

The latest scheduled dates are used for the management review. On the above tabulation, the events shown with float time (and therefore not critical) could be expected to be completed earlier by the amount of the float times shown in the tabulation (if all goes according to plan). The dates shown are the latest scheduled time to complete the event. For the report, it is prudent to show this date. The event can "drift" within the float, without a need for any explanation. Explanatory remarks should be minimal, but in this case, they will be needed to show why specific activities within the *Building* division are now expected to be completed on May 6, 1996 instead of the original April 22, 1996 completion date.

USING THE COMPUTER

Once the concept of milestones is understood, preparing a computerized milestone report is a more desirable approach than the time-consuming manual method. To simplify data entry, we suggest adding the milestone data to Figure 4.9, the activities/resources worksheet. Place the node number of the milestone

Activities/Resources Worksheet

Project Name:
Christopher Design/Build Project

Start Node	End Node	Duration (Days)	Description	Resources (Daily Requirements)			
				EN	AR	DE	SP
Start	1	—	_____	—	—	—	—
1	2	14	Design structural steel	2	—	6	—
1	3	91	Design building	6	4	12	—
1	4	42	Design mechanical equipment	4	—	4	—
2	3	70	Fabricate and erect structural steel	1	—	2	10
3	7	42	Construct phase I building	3	—	—	20
4	5	28	Design electrical equipment	4	—	8	—
4	6	63	Fabricate mechanical equipment	2	—	4	—
4	8	14	Design interior items	—	4	8	—
5	6	49	Procure electrical equipment	2	—	1	—
6	7	42	Install mechanical and electrical equipment	4	—	—	16
7	11	42	Construct phase II building	5	—	—	20
8	9	28	Procure interior items	—	2	2	—
9	10	70	Install interior items	—	2	—	12
10	11	—	Dummy	—	—	—	—
11	F	—	_____				
1	Milestone		Start project				
3	Milestone		Comlete structural steel				
6	Milestone		Complete electrical equipment–				
			design and procure				
7	Milestone		Complete mechanical and electrical;				
			Complete phase I				
10	Milestone		Complete architectural				
11	Milestone		Complete project				

EN—Engineers
AR—Architects
DE—Designers
SP—Skilled Personnel/Labor

FIGURE 4.9   Activities/resources worksheet for the Christopher Design/Build Project

under the Start Node, type M (for MILESTONE) under the End Node, then complete the DESCRIPTION column.

To enter milestone data into the computer, bring up the Christopher Design/ Build Project by pressing 1: Produce an Up-to-Date Plan? Select the DGP version and then press F3 to get a blank input screen.

1. At line 1, type in the From column the node number shown in the Start column of the Worksheet [ENTER].
2. Type **M** under **To** [ENTER]. (The word Milestone will be printed on the screen.)
3. Type the description of the milestone [ENTER].
4. Type in the descriptions of the remaining milestones.

If you had started with a blank screen, the completed milestone screen would appear like this:

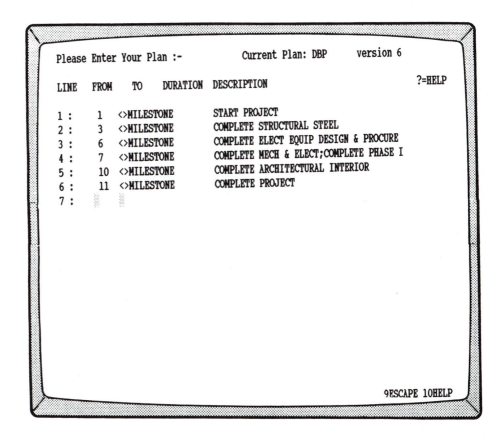

Press [ESC] to have milestones listed with the rest of the plan. The first screen with the milestone data will look like this:

```
COMMAND?                        Current Plan: DBP      version 6

LINE   FROM    TO    DURATION  DESCRIPTION                        ?=HELP

 1 :  START-> 1               NOT BEFORE DAY     2
 2 :   1    <>MILESTONE       START PROJECT
 3 :        -> 2        14    DESIGN STRUCTURAL STEEL
 4 :        -> 3        91    DESIGN BUILDING
 5 :        -> 4        42    DESIGN MECHANICAL EQUIPMENT
 6 :   2    -> 3        70    FABRICATE & ERECT STRUCTURAL STEEL
 7 :   3    <>MILESTONE       COMPLETE STRUCTURAL STEEL
 8 :        -> 7        42    CONSTRUCT PHASE I BUILDING
 9 :   4    -> 5        28    DESIGN ELECTRICAL EQUIPMENT
10:        -> 6        63    FABRICATE MECHANICAL EQUIPMENT
11:        -> 8        14    DESIGN INTERIOR ITEMS
12:   5    -> 6        49    PROCURE ELECTRICAL EQUIPMENT
13:   6    <>MILESTONE       COMPLETE ELECT EQUIP DESIGN & PROCURE
14:        -> 7        42    INSTALL MECH & ELECT EQUIP
15:   7    <>MILESTONE       COMPLETE MECH & ELECT;COMPLETE PHASE I
16:        -> 11       42    CONSTRUCT PHASE II BUILDING
17:   8    -> 9        28    PROCURE INTERIOR ITEMS
18:   9    -> 10       70    INSTALL INTERIOR ITEMS
19:  10    <>MILESTONE       COMPLETE ARCHITECTURAL INTERIOR

1DESCRP 2RESRCE 3INSERT 4NEW PG 5HDINGS 6GO TOP 7ZOOM   8BARCHT 9ESCAPE 10HELP
```

To get a milestone report, bring up the menu showing the list of options for project reporting, and press **1**: Standard Report Listing:

1. Press **1**: The Earliest Activities First?
2. Type **Milestone** [ENTER].
3. Press [ENTER] at (2) *Starting Location*.
4. Press **8**: Start or Cancel the Run [ENTER]. (See Figure 4.10.)

The advantage of this report is its summarizing value. Instead of examining the status of every activity, the reader can review a summary of the project with the milestone report:

1. The report confirms the start of the project (2/OCT/95) and its completion (22/APR/96).
2. There is a time span (available float) for the event, *Complete structural steel*—1/JAN/96 to 29/JAN/96—that will not affect the project completion date.
3. The events, *Complete electrical equipment* and *Complete mechanical and electrical equipment*, are critical—they must be completed on 29/JAN/96 and 11/MAR/96, respectively.

```
PROJECT MANAGEMENT                          DESIGN/BUILD PROJECT                              BOCA RATON, FLORIDA
        REPORT TYPE :MILESTONE                                         PRINTING SEQUENCE  :Earliest Activities First
                                                                       SELECTION CRITERIA :MILESTONE
        PLAN I.D.    :DBP     VERSION  1                               TIME NOW DATE       : 2/OCT/95
===============================================================================================================
            MILESTONE                       EARLIEST EVENT             LATEST EVENT
            DESCRIPTION                      TIME                      TIME
===============================================================================================================
    1-   1 START PROJECT                     2/OCT/95                  2/OCT/95
    3-   3 COMPLETE STRUCTURAL STEEL         1/JAN/96                  29/JAN/96
    6-   6 COMPLETE ELECTRICAL EQUIPMENT     29/JAN/96                 29/JAN/96
   10-  10 COMPLETE (ARCHITECTURAL) INTERIOR 4/MAR/96                  22/APR/96
---------------------------------------------------------------------------------------------------------------
    7-   7 COMPLETE MECH & ELECT;COMP PHASE II 11/MAR/96               11/MAR/96
   11-  11 COMPLETE PROJECT                  22/APR/96                 22/APR/96
===============================================================================================================
```

FIGURE 4.10   Milestone report for the Christopher Design/Build Project

> 4. There is available float for the event, *Complete architectural interior*, which can be completed from 4/MAR/96 to 22/APR/96.

The summary bar chart for the Christopher Design/Build Project (Figure 4.7) will reflect some delays and will be compared with the last report to note any adverse trends. Affected events will be investigated further for possible corrections.

The milestone report is usually added to the project status report, since it supplements the timing data being submitted. Milestones essentially summarize the status of the project.

The computerized milestone report has certain benefits:

- The milestone listing can be entered in the data bank at the same time as the project activities, thereby saving time for other work.
- Additional milestones can be added during the course of the project with minimal time and effort.
- Status report changes can be effected quickly and automatically. Changes to the project activities are transferred directly to the milestones during the data inputting process.
- The computer program can sort the critical events requiring attention, highlighting activities that lead up to the milestone.

PROJECT REPORTS

During the term of the project, an inclusive report may be needed that reviews the overall progress to date or a report may be prepared at the end of the project for documentation. As status reports described in this chapter are relatively short for such purposes, a conventional type of report used in the business environment may be more appropriate. Key elements of this type of report include:

    I. Introduction
      A. Purpose
      B. Scope (limits of report)
      C. Background

II. Summary (highlights of project)
III. Conclusions
IV. Project description
V. Appendix

The introduction should describe the purpose (a statement of the objective of the report), scope (limitations associated with the concept, project objectives, purpose, timing, costs, personnel/labor), and background (facts behind the project).

The summary (which may also be called Highlights or Executive Highlights) can be a commentary or made up of numbered or bulleted statements. Its contents summarize the completed project report. If the report is relatively small, then the summary may also contain the decisions and judgments stemming from the project's outcome.

The conclusions may also contain any recommendations. In fact, each conclusion may be followed with a recommendation. This section may also contain possible alternative courses of action that could or should have been taken. If the report is written at the beginning of the project, it should include a rationale for the preferred course of action.

The remainder of the report, project description and appendix, contains the details (work breakdown structure, planning diagram, personnel/labor requirements, costs, etc.), calculations, and reference material to support the introduction, summary, conclusions, and recommendations. The project description represents the full-scale report—what was done, how it was done, and why it was done. Although the introduction with the summary and conclusions are more widely read, the voluminous detail in the description, nonetheless, should be accurate and comprehensive. It is the substantiation of the conclusions.

A typically structured report will follow this outline:

Title Page

A. Letter of transmittal
B. Table of contents
C. Introduction
D. Summary
E. Conclusions
F. Project description
    1. Plan
        a. Objective statement(s)
        b. List of project activities (with descriptions)
        c. Work breakdown structure
        d. Planning diagram
    2. Schedule
        a. Time estimates
        b. Calendar schedule
        c. Bar chart schedule

3. Project control
    a. Status reports
    b. Progress bar chart schedules
    c. Summary bar charts
4. Project costs
    a. Cost (expenditures) schedule
    b. Cost control (indicated cost outcome reports)
    c. Time/cost tradeoffs (where applicable)
5. Personnel/labor plan and schedule
    a. Load distribution charts of personnel/labor
    b. "Smoothing" charts of personnel/labor
G. Appendix
    1. Calculations
        a. Subdiagrams (planning diagram worksheets)
        b. Planning diagram data (shown on diagram: earliest start and latest finish times, critical path)
        c. Total float tabulation
        d. Cost tabulation
        e. Personnel/labor tabulations
    2. Computer printouts
        a. Timing schedules
            (1) Early start
            (2) Total float
            (3) Bar charts
        b. Cost schedules
        c. Labor/personnel schedules
    3. Reference material

SUMMARY

Project control measures the performance of the project by tracking its performance. Maintaining a systematic pattern throughout the project is essential to achieving successful results. The necessary actions to be taken follow a four-step early warning system procedure:

1. monitoring the project activities through periodic checking of their time deviations
2. assessing the project activities by identifying adverse trends and critical activities
3. resolving any concerns after assessing the critical activities
4. communicating status findings to project participants, vendors, suppliers, contractors, and management through phone calls, memos, meetings, and periodic written reports

An effective means of communication used to inform participants of project performance is the project status report. Along with a letter of transmittal, it will include a bar chart progress schedule based upon the calculated time schedules. This is a graphic report given to management to supplement the commentary in the status report; together, they assist in a sound performance appraisal of the project. The report can also review performance of vendors, suppliers, and contractors. The status report highlights the critical areas and allows objective evaluation of the problem areas.

A milestone schedule report is another project control expedient that helps enhance use of the status report. Milestones are selected events that are of major importance in achieving objectives.

Using a computer in the project control phase has many advantages:

- Its data bank stores previously entered information about the progress of the project that can be updated quickly.
- Bar charts can be generated promptly for analyzing alternate courses of action if changes to the project are required.
- Computer-tabulated schedules coordinated with the bar charts showing optional courses of action are most valuable in resolving project problems.

Exercises

1. Identify the four-stage procedure to effectively control a project. Briefly describe each stage with one or two sentences.
2. Using computer software, prepare a progress schedule for the project whose plan is shown in Figure 1.2 and using the original schedule developed in Chapter 3 Exercises 3 and 4. Use the following information to prepare the progress schedule:
   a. The progress schedule should consist of tabulated schedule, bar chart, and milestone report.
   b. Reporting date is week 15.
   c. At the monitoring stage, the project team determined that the durations of five activities have changed from their original estimates:

	Revised Duration (Weeks)
Fabricate and erect structural steel	12
Phase II construction	15
Equipment fabrication and delivery	14
Long-lead items—engineering	10
Long-lead items—procurement	24

3. Refer to Exercise 2 and answer the following questions that the letter accompanying the status report will need to address:
   a. What is the new project completion date?
   b. List the critical items, durations, and their new latest finish dates.
4. Refer to Exercise 2: The project duration has changed from _____ weeks to _____ weeks. To help achieve the initial objective of meeting the original completion date, select one activity whose duration may be reduced, and by how much.
5. Monitoring the Christopher Design/Build Project at week 12 (25/Dec/95) showed the following activities had deviated from their original schedule:

Activity	Duration (Weeks)	
	Original	Revised
Design structural steel	2	4
Design building	13	16
Design electrical equipment	4	6
Procure interior items	4	8

Prepare a milestone schedule reflecting the above revised durations.

CHAPTER 5

Project Costs: Schedule and Control

THIS CHAPTER WILL COVER

- Procedure for scheduling costs

- Procedure for controlling costs

- Using the computer for scheduling and controlling costs

Previous chapters covered methods for planning, scheduling, and controlling project timing in an organized manner to achieve successful results. Just as important to a project's success are procedures to plan, schedule, and control project costs.

Those involved in the project pay a great deal of attention to budgets and cash flows (having adequate revenues to handle expenses) from the planning stage to the end of the project. The impact of costs (and timing, which is closely related) on the outcome of the project is a major consideration. Those controlling the project funds will maintain the financial integrity of the project by keeping two cost particulars in control:

1. *What* amounts of money will be needed at select times during the course of the project?
2. *When* will the money be needed to pay for materials, labor, and other expenses over the course of the project?

Controlling these two factors continuously during the project is essential. This chapter examines several fundamental cost factors that are helpful to meet the budget and to keep a "healthy" cash flow:

- scheduling project costs to know when payments are due
- controlling project costs by predicting the final cost outcome in earlier periods of the project
- minimizing costs by determining the minimum additional costs needed to speed up a project (sometimes referred to as time/cost trade-offs)

While these techniques are not necessarily the only ones used in project cost control, they are among the most important ones to help achieve financial stability.

SCHEDULING PROJECT COSTS

The project cost schedule uses the budget as a guide to distribute money among the project activities. The budget, one of the major cost objectives set up at the beginning of the project, is usually a "lump sum" figure. It has insufficient detail to identify the individual costs incurred by completing each activity. The process of dividing the budget into activity costs can be termed the cost planning phase.

Planning Project Costs

Project costs are planned at the beginning of the project. Identifying costs for completing each activity and preparing timing estimates have similar patterns. Those knowledgeable of and/or responsible for completing specific project activities should provide the estimates. There are other alternatives if this presents a problem.

Supplier quotations and contractor proposals are sources of estimates. Many firms will have on file historical costs that they draw on for new estimates. If none of these are available, they may need to make educated guesses on the project activities. Whatever the source, it is important to use the best possible estimates as these costs may become rigid and become the standard throughout the project. Adding reasonable contingencies to selected cost items may be necessary where uncertainties exist about completing activities.

Contingencies (allowing additional costs for potential problems such as added work, potential design changes, work stoppages, etc.) are incorporated in the costs. The amount of the contingency is linked to the risk associated with completing the specific activity. While they will probably not make it known, those providing estimates who are also responsible for the specific activities will generally include a contingency. They justify this action to cover uncertainty.

The costs finally developed for the Christopher Design/Build Project are as follows:

Activity	Estimated Cost
Structural steel	$ 40,000
Design building	420,000
Design mechanical equipment	120,000
Fabricate and erect structural steel	220,000
Construct phase I building	420,000
Design electrical equipment	80,000
Fabricate mechanical equipment	90,000
Design interior items	40,000
Procure electrical equipment	40,000
Install mechanical and electrical	200,000
Construct phase II building	450,000
Procure interior items	40,000
Install interior items	240,000

DEVELOPING THE COST SCHEDULE

The cost schedule is necessary to carry out the cost objectives of the project.

- It is necessary to know when money is needed and how much money is needed at all times during the project.
- The prepared cost schedule is the basis for comparing the project costs with actual spending.
- The cost schedule is necessary for depreciation schedules.
- For building projects, the project cost schedule together with the project time schedule become the basis for determining when various property tax payments become due.

Developing the project cost schedule will follow these steps:

1. Use the bar chart time schedule as the initial document.
2. On the bar chart time schedule, place the costs needed to complete each project activity under the description of their respective project activities.
3. The bar chart cost schedule is formed when all estimates are placed on the bar chart time schedule.
4. Prepare a tabulation format for the cost/time unit. Its main function is to ease the calculations necessary for the cost scheduling tabulations. The **cost/time unit** is the cost incurred in performing the project activity over a specified unit of time. Depending on the conditions, you may prepare it on a weekly basis (i.e., it would be shown as dollars/week; monthly, dollars/month). Computer scheduling programs may define it on a daily basis.
5. Before preparing the cost scheduling format, determine the time increments at which the costs are to be considered. As most status reporting is done monthly, this is the most popular frequency.
6. Calculate the total costs of each activity within each time period. While the one-month period normally is used to coincide with the status reporting period, shorter periods of time (bimonthly, weekly) may also be used to control costs. This will normally imply that the cost status of the project is becoming critical and more frequent cost reviews are desired. Conversely, longer reporting periods suggest spending is proceeding as expected or better than expected. The Christopher Design/ Build Project uses a five-week period to coincide with its five-week report status period.

Bar Chart Cost Schedule

Prerequisite to a project cost schedule is a completed bar chart timing schedule. It shows each activity and the cost to complete that activity. The sum of all of these activities' costs equals the total cost to complete the project. The cost to complete each activity for the Christopher Design/Build Project is depicted in a bar chart in Figure 5.1.

The costs identified with project activities shown on the bar chart cost schedule are known as variable costs. Not shown are fixed costs related to the project, which include overhead, insurance, office rentals, office expenses, and field supervision costs. They will be discussed later in the chapter when cost minimizing is addressed.

Calculating the Cost/Time Unit

The cost/time unit is the cost of each project activity over a unit length of time. It is developed to simplify cost schedule calculations. Once the unit of time is selected (one week for the Christopher Design/Build Project), a format for

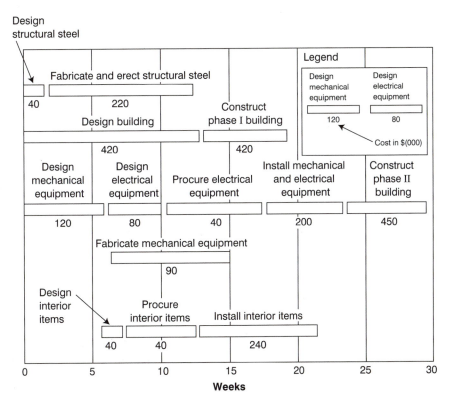

FIGURE 5.1 Bar chart cost schedule for the Christopher Design/Build Project

developing the cost/time unit is produced. Instructions to construct the cost/time unit format are as follows:

Cost/Time Unit

Activity	Total Cost ($)	Duration (Weeks)	Cost/Time (Dollars/Week)
(1)	(2)	(3)	(4)

(1) Activity	Enter the description of the activity.
(2) Total cost	Enter the total estimated cost in dollars of performing the project activity.
(3) Duration	Enter the total duration time of the project activity.
(4) Cost/time	Enter the result of dividing column 2 by column 3.

Figure 5.2 shows the cost/time unit format for the Christopher Design/Build Project.

Project Activity	Cost	Duration (Weeks)	Cost/Week
Design structural steel	$ 40,000.00	2	$20,000.00
Design building	420,000.00	13	40,000.00
Design mechanical equipment	120,000.00	6	20,000.00
Fabricate and erect structural steel	220,000.00	10	22,000.00
Construct phase I building	420,000.00	6	70,000.00
Design electrical equipment	80,000.00	4	20,000.00
Fabricate mechanical equipment	90,000.00	9	10,000.00
Design interior items	40,000.00	2	20,000.00
Procure electrical equipment	40,000.00	7	5,714.00
Install mechanical and electrical equipment	200,000.00	6	33,333.00
Construct phase II building	450,000.00	6	75,000.00
Procure interior items	40,000.00	4	10,000.00
Install interior items	240,000.00	10	24,000.00

FIGURE 5.2   Calculating the cost/time format for the Christopher Design/Build Project

Calculating the Cost Schedule

The cost scheduling process, summarized in the cost schedule tabulation, is used as a basis for preparing the periodic status of project expenditures. Before starting the cost schedule, complete the cost/time unit tabulation and the bar chart cost schedule. Specific instructions for developing the cost schedule format are as follows:

Cost Schedule

Time Period	Activity	Duration (Weeks)	Cost/Time (Dollars/Week)	Expenditures ($)
(1)	(2)	(3)	(4)	(5)

(1) Time period   Enter the time period selected. (Remains constant for the total tabulation. Time period is usually the same period used for status reporting.)

(2) Activity   Enter the description of the project activity.

(3) Activity time   Enter the amount of time that the activity performs within the time period.

(4) Cost/time   Enter the cost/time of the activity.

(5) Expenditures   Enter the result of multiplying column 3 by column 4.

This tabulation shows the planned budget for every five weeks—the period used for cost status reporting on the Christopher Design/Build Project. The period should not change unless management requests it or conditions change. The cash flow of the project may be critical, for example, and management needs to know what costs will be incurred over shorter incremental periods. The sum of the expenditures for every period will equal the total cost of the project. This type of schedule allows comparison of the differences between actual and planned project costs for any given period.

Figure 5.3 shows the cost schedule format for the Christopher Design/Build Project.

Constructing the Cost Distribution Graph

The cost distribution graph is a graphic portrayal of the cost schedule. Combined with the bar chart cost schedule, it provides a good understanding of the amount of expenditures needed at any given time. A good method for constructing the cost distribution graph is to use the bar chart cost schedule as a template to lay out the graph in proper time increments. Note that Figure 5.4(a), the bar chart cost schedule, is plotted above Figure 5.4(b), the cost distribution graph for the Christopher Design/Build Project. (The values used are for a five-week cost schedule.)

A picture of the cost schedule in this graphic format can influence how the costs are distributed over the project, showing peaks and valleys in the spending schedule. For example, this graph points out that spending is higher in the first part of the project than in the latter part, which may not be desirable. Let us assume that the contractor is being compensated every month based on the amount of work completed. The contractor may need a temporary bank loan if the expenses exceed compensation received for the work being performed. A more desirable alternative is arranging the project work in such a way that the incoming revenue exceeds expenses. In this particular project, a good business maneuver would be to shift some of the costs to the latter part of the project.

Adjusting the Cost Schedule

Upon reviewing the cost distribution graph for the Christopher Design/Build Project, the contractor has decided that costs incurred in the period between weeks 16 and 20 are more than the funds that can be made available for that period. In the business world, this is known as a period of tight cash flow because of a high outflow of expenses and an inadequate revenue inflow.

To ease this situation, the existing cost schedule (developed with the early start time schedule) can be adjusted. Investigating the late start schedule (using

Period	Project Activity (Description)	Activity Time (Weeks)	Unit Cost (Per Week)	Total Activity Costs
0–5	Design structural steel	2	$20,000	$ 40,000
	Design building	5	40,000	200,000
	Fabricate and erect structural steel	3	22,000	66,000
	Design mechanical equipment	5	20,000	100,000
	Total 0–5			406,000
6–10	Design building	5	40,000	200,000
	Fabricate and erect structural steel	5	22,000	11,000
	Design mechanical equipment	1	20,000	20,000
	Design electrical equipment	4	20,000	80,000
	Fabricate mechanical equipment	4	10,000	40,000
	Design interior items	2	20,000	40,000
	Procure interior items	2	10,000	20,000
	Total 6–10			510,000
11–15	Design building	3	40,000	120,000
	Fabricate and erect structural steel	2	22,000	44,000
	Construct phase I building	2	70,000	140,000
	Procure electrical equipment	5	5,714	28,571
	Fabricate mechanical equipment	5	10,000	50,000
	Procure interior items	2	10,000	20,000
	Install interior items	3	24,000	72,000
	Total 11–15			474,571
16–20	Construct phase I building	4	70,000	280,000
	Procure electrical equipment	2	5,714	11,428
	Install mechanical and electrical equipment	3	33,333	99,999
	Install interior items	5	24,000	120,000
	Total 16–20			511,427
21–25	Install mechanical and electrical equipment	3	33,333	99,999
	Construct phase II building	2	75,000	150,000
	Install interior items	2	24,000	48,000
	Total 21–25			297,999
26–29	Construct phase II building	4	75,000	300,000
	Total Project Costs			$2,500,000*
	*Rounded.			

FIGURE 5.3   Cost schedule for the Christopher Design/Build Project

(a)

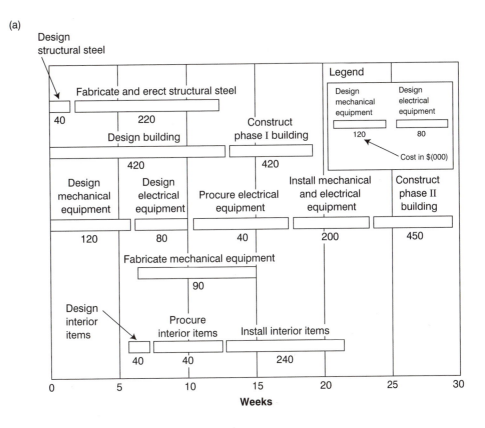

(b)

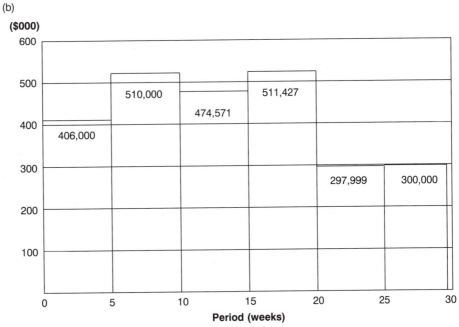

FIGURE 5.4 (a) Bar chart cost schedule; (b) cost distribution graph for the Christopher Design/Build Project

the same procedure as the early start schedule) shows that the cash flow can be relieved without jeopardizing the project completion date.

A helpful suggestion to simplify preparing the late start cost distribution graph is to identify the float times of the project activities on the bar chart cost schedule. At times this can provide an early idea of whether the late start schedule has adequate opportunity for necessary adjustments.

For example, when examining possible candidates for adjustments in the Christopher Design/Build Project, *Install interior items* stands out because of its high cost and large float value. By pushing this project item to its latest start/ latest finish time (which will not affect the project completion time), this not only helps yield a more uniform cash flow but permits much of the interior work to be completed near the end of the project, thereby avoiding possible damage from other work areas.

This change with the revised five-week schedule compared to the original schedule is shown as follows:

| | Total Costs | |
Weeks	Original	Revised
0–5	$406,000	$406,000
6–10	510,000	510,000
11–15	474,571	402,571
16–20	511,427	415,427
21–25	297,999	369,999
26–29	300,000	396,000

Figure 5.5 shows the revised cost schedule calculations, reflecting the change in the time schedule of *Install interior items*. Figure 5.6 shows the revised bar chart cost schedule, plotting the *Install interior items* bar with its latest start/ latest finish, and the corresponding revised cost distribution graph for the Christopher Design/Build Project. Its "smoother" cost distribution is accented when shown graphically.

This procedure will be used during the course of the project if critical project activities with relatively high costs appear to be deviating from the original schedule. Those controlling project funds will closely observe how they affect cash flow during the rest of the project.

PROJECT COST CONTROL

In the previous chapter on project control we discussed project timing status and using a project status report as the principal means of communicating. This

Period	Project Activity (Description)	Activity Time (Weeks)	Unit Cost (Per Week)	Total Activity Costs
11–15	Design building	3	$40,000.00	$ 120,000.00
	Fabricate and erect structural steel	2	22,000.00	44,000.00
	Construct phase I building	2	70,000.00	140,000.00
	Procure electrical equipment	5	5,714.00	28,571.00
	Fabricate mechanical equipment	5	10,000.00	50,000.00
	Procure interior items	2	10,000.00	20,000.00
	Total 11–15			402,571.00
16–20	Construct phase I building	4	70,000.00	280,000.00
	Procure electrical equipment	2	5,714.00	11,428.00
	Install mechanical and electrical equipment	3	33,333.00	99,999.00
	Install interior items	1	24,000.00	24,000.00
	Total 16–20			415,427.00
21–25	Install mechanical and electrical equipment	3	33,333.00	99,999.00
	Construct phase II building	2	75,000.00	150,000.00
	Install interior items	5	24,000.00	120,000.00
	Total 21–25			369,999.00
26–29	Construct phase II building	4	75,000.00	300,000.00
	Install interior items	4	24,000.00	96,000.00
	Total 26–29			$ 396,000.00

FIGURE 5.5   Revised cost schedule for the Christopher Design/Build Project

is the same type of reporting done for the status of project costs. In fact, it may become part of the project timing status report. (To repeat a constructive statement: Project timing and costs are the very essence of project management. Combining their status in one report is good project reporting in many cases.) While cost and time summaries may be combined for reporting purposes, investigating detailed costs status takes on a different procedure than when investigating the timing status.

The basis of the project cost status report is the indicated cost outcome report (see Figure 5.7). It is an "early warning" system designed to alert all project participants of any potential adverse cost variations early enough in the program to consider corrections. The main purpose of the indicated cost outcome report is to ensure that project spending is contained within the approved budget. It compares the authorized spending (budget) with the actual spending projected to the end of the project. The actual spending is equal to spending to date plus the projected future spending.

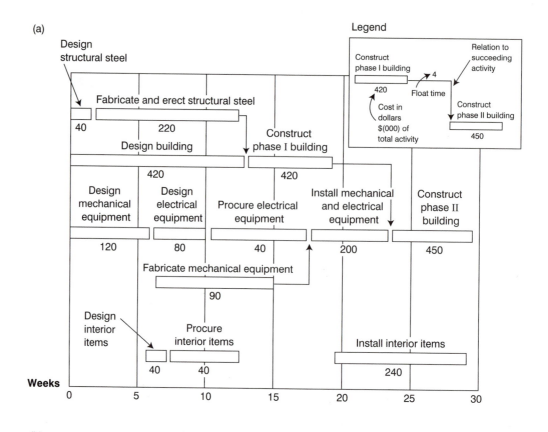

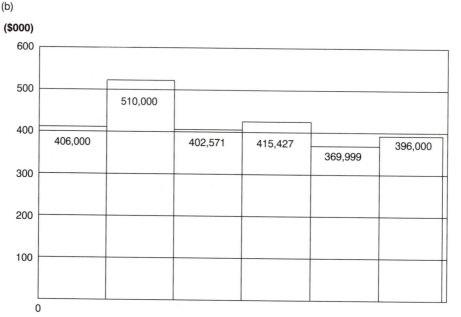

FIGURE 5.6 (a) Revised bar chart cost schedule; (b) revised cost distribution graph for the Christopher Design/Build Project

Project Activity (Description)	Authorized	Committed (Spent) to Date	Future Commitments	Indicated Outcome	Variance (Over)/Under	Percentage Variance (Over)/Under
Design structural steel	$ 40,000.00	$ 44,000.00	0.00	$ 44,000.00	$ (4,000.00)	(10.0)%
Design building	420,000.00	420,000.00	0.00	420,000.00	0.00	0.0
Design mechanical equipment	120,000.00	110,400.00	0.00	110,400.00	9,600.00	8.0
Fabricate and erect structural steel	220,000.00	244,420.00	0.00	244,420.00	(24,420.00)	(11.1)
Construct phase I building	420,000.00	300,000.00	78,000.00	378,000.00	42,000.00	10.0
Design electrical equipment	80,000.00	60,500.00	14,460.00	74,960.00	5,040.00	6.3
Fabricate mechanical equipment	90,000.00	70,500.00	12,570.00	83,070.00	6,930.00	7.7
Design interior items	40,000.00	20,000.00	20,000.00	40,000.00	0.00	0.0
Procure electrical equipment	40,000.00	35,000.00	10,720.00	45,720.00	(5,720.00)	(14.3)
Install mechanical and electrical equipment	200,000.00	10,000.00	230,000.00	240,000.00	(40,000.00)	(20.0)
Construct phase II building	450,000.00	200,000.00	317,500.00	517,500.00	(67,500.00)	(15.0)
Procure interior items	40,000.00	30,000.00	6,680.00	36,680.00	3,320.00	8.3
Install interior items	240,000.00	80,000.00	226,480.00	306,480.00	(66,480.00)	(27.7)
	2,400,000.00	1,624,820.00	916,410.00	2,541,230.00	(141,230.00)	(5.9)

FIGURE 5.7 Indicated cost outcome report (week 15) for the Christopher Design/Build Project

The base document of the indicated cost outcome is a tabulation format. Instructions to complete the format are as follows:

Column	Description	Procedure
1	Project item	Enter the description with each project activity.
2	Authorized budget	Enter the estimated dollar amount that has been developed and approved for completing the project activity item.
3	Committed to date	Enter for each activity the amount that has been spent and/or ordered as of the date of the report.
4	Future commitments	Enter the additional costs that will be needed to complete the activity. These projections are usually estimated.
5	Indicated outcome	Enter the sum of columns 3 and 4 for each activity. Compare this figure to the authorized budget to determine the cost performance of each item. All the indicated outcomes of activities are added algebraically, and this total is compared with the total budget cost.
6	Variance (over) or under	Enter the difference between columns 2 and 5. When the difference is over the authorized amount (known as the overrun) beyond a predescribed tolerance (usually 10%) of the authorized amount, project operations may be obliged to stop until there is assurance that no further overruns can be expected.
7	Percentage (%) of variance (over) or under	Enter the result of dividing the difference between column 2 and column 5, divided by column 2. (Multiply by 100 to arrive at a percentage.)

Five items fit this category, and their review, analysis, and resolutions are shown in Figure 5.8. These highlights supplement the indicated cost outcome report and are usually attached to the report to emphasize the cost concerns and show that action is being taken to resolve the problems.

One of the major concerns in this project is the cost of the electrical equipment items, which needs to be reduced. The project team reviewed and analyzed the electrical equipment cost items with their vendors and suppliers, then met with the building construction and equipment installation contractors. Their efforts to have the costs reduced and bring the project within budget were successful.

CHRISTOPHER DESIGN/BUILD PROJECT

Reporting Date: Week 15

Indicated Cost Outcome Report
Executive Highlights

- Project is 50% complete
- Project expenditures (commitments) to date: 67% of authorized (budgeted) amount
- Project expenditures are $141,230 (5.9%) over authorized budget
- Outstanding project activities with highest overruns:

Project Activity	Overrun	
	Variance ($)	Variance (%)
Fabricate and erect structural steel	(24,420)	(11.1)
Procure electrical equipment	(5,720)	(14.3)
Install mechanical and electrical equipment	(40,000)	(20.0)
Construct phase II building	(67,500)	(15.0)
Install interior items	(66,480)	(27.7)
	(204,120)	

- Recommendations to reduce overruns:

Project Activity	Recommendations	
1. Fabricate and erect structural steel	Work is essentially complete. Very little opportunity to reduce overrun.	$ 0.00
2. Procure electrical equipment	Same situation as structural steel.	0.00
3. Install mechanical and electrical equipment	Defer installation of nonessential equipment until after project is complete.	(23,000)
4. Construct phase II building	Revise bid package to show reduction in scope of work.	(70,000)
5. Install interior items	Defer nonessential work until after project is complete.	(50,000)
	Total variance reduction:	($ 143,000)

- Project team developing details on variance reduction recommendations. Status to be included with next report.

FIGURE 5.8 Executive highlights for the Christopher Design/Build Project

If they had not succeeded in reducing the cost of the electrical items, there would be a review of activities that do not show overruns, but may have the potential of lower costs. Cost reviews are continuous, with many projects facing endless cost overruns. The actions needed depend on the severity of the situation.

Snyder Engineering insists on sticking to the authorized budget for this project and correcting any projected cost overrun. Program costs must be contained, and those responsible for the project funds may elect to refuse authorization of any additional expenditures to pay for the overruns. This, in effect, will shut down the project, so every effort must be made to bring the budget under control.

Although this may seen harsh, the results would be more disastrous if these potential overruns were not identified and acted upon as early as possible. Shutting down the project in an early phase is not necessarily an arbitrary decision. It is a "wake-up call" for the project participants to examine the critical situation and make the necessary adjustments to ensure that when operations begin, funds are available to complete the project.

Some firms may allow funds for overruns on projects, maybe as much as 10 percent of the total authorized amount. This may happen if the cost estimates for the project were prepared with limited information and a high degree of uncertainty. Conceptual cost estimates—those developed from incomplete designs—may have provisions for added funds if the actual costs exceed estimated costs.

COST MINIMIZING

Snyder Engineering wishes to explore the cost impact of completing the Christopher Design/Build Project earlier than the present 29 weeks. To do so, the firm will use the cost minimizing (sometimes called the time/cost tradeoffs) method.

Cost minimizing in project management involves determining how to reduce the time required for completing a project with the fewest added costs. Reducing the duration of a project usually means overtime for labor/personnel, additional labor/personnel, and additional equipment—all implying additional funds. While Snyder Engineering wants to avoid any situation that will result in added costs to the project, the impact of reducing the schedule deserves investigation, since occupying the building earlier may afford additional revenues to the owner. It is good business sense and Snyder Engineering wants the owner to consider this possibility, if indeed it does exist.

Determining the potential cost increases of reducing a project's duration requires estimates of two factors for each project activity:

1. the normal cost required for accomplishing the work on a normal time basis
2. the additional cost required for accomplishing the work on an expedited basis (and, as a basis for this additional expenditure, the corresponding reduction in time that the work requires)

Using the two sets of estimates, alternative schedules are developed and used to determine the additional costs, if any, to achieve the best reduced project completion time. ("If any" in the previous sentence suggests that other factors may enter into the equation that actually lower costs.)

The cost minimizing procedure follows the "cut and try" pattern and can be done with manual calculations. Basic documents needed include the Christopher Design/Build Project planning diagram and the timing schedule that have already been developed. For ease in making the calculations for the cost minimizing schedule they are shown again in Figure 5.9. Also refer to the cost schedule already developed in Figure 5.1.

(a)

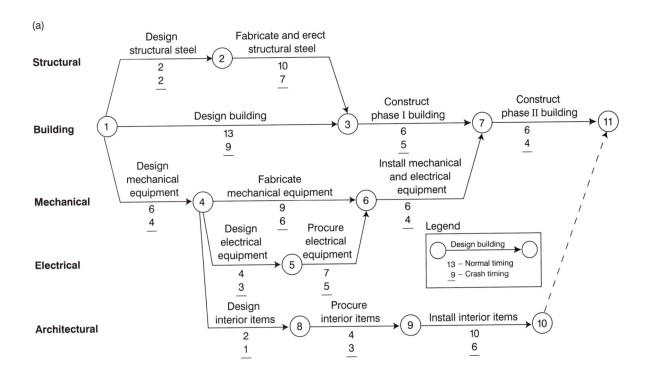

(b)

i-j	Description	Time (Weeks)	Earliest		Latest		Total Float
			Start	Finish	Start	Finish	
1-2	Design structural steel	2	0	2	5	7	5
1-3	Design building	13	0	13	4	17	4
1-4	Design mechanical equipment	6	0	6	0	4	0
2-3	Fabricate and erect structural steel	10	2	12	7	17	5
3-7	Construct phase I building	6	13	19	17	23	4
4-5	Design electrical equipment	4	6	10	6	10	0
4-6	Fabricate mechanical equipment	9	6	15	8	17	2
4-8	Design interior items	2	6	8	13	15	7
5-6	Procure electrical equipment	7	10	17	10	17	0
6-7	Install mechanical and electrical equipment	6	17	23	17	23	0
7-11	Construct phase II building	6	23	29	23	29	0
8-9	Procure interior items	4	8	12	15	19	7
9-10	Install interior items	10	12	22	19	29	7
10-11	Dummy	0	22	22	29	29	7

FIGURE 5.9 (a) Project planning diagram; (b) time schedule for the Christopher Design/Build Project

The first step is to tabulate the normal time and cost for each activity from the planning diagram and the time and cost schedules. The next step is to develop the crash time and crash costs for each activity:

1. Crash time is the minimum estimated time in which a project activity could be completed if the job is accelerated.
2. Crash cost is the normal cost plus the extra cost involved in accelerating an activity.

Crash time and crash cost estimates for the activities will be acquired from those responsible for or knowledgeable of the specific project activities. They should answer these questions:

- Are you able to accelerate your job, and by how much time?
- Will any additional costs be incurred? Show the added costs incrementally (i.e., how much for each week that your activity is reduced).
- Of these additional costs, how much will be for overtime, added personnel/labor, and/or added equipment?
- Will any other costs be incurred?

Crash time and crash cost for each activity are needed to determine the added cost. The formula used to calculate the added cost per unit of reduced project time for each project activity is as follows:

$$\text{added unit cost/unit time} = \frac{\text{crash cost} - \text{normal cost}}{\text{normal time} - \text{crash time}}$$

Certain assumptions are necessary in using the cost-minimizing method to accelerate a project. The first is that crash time is always less than or equal to normal time. The second is that the time/cost curve is generally linear (a straight line allows for more accuracy).

These two points are illustrated in Figure 5.10, which shows the time/cost curve for *Design mechanical equipment*.

When all the project activities are analyzed for their crashing capabilities, a complete tabulation is made of all of the project activities (Figure 5.11) showing activity descriptions; critical activities, which are specifically noted; normal and crash costs of each activity; normal and crash timing of each activity; and the added cost/unit time value needed to shorten the duration of the activity.

Once this tabulation is complete, the next step is to determine the total minimum additional cost for reducing the project duration. Snyder Engineering has decided to show the effect of the additional project costs when reducing the project duration in two-week increments, that is, the minimum additional cost if the project duration is 27 weeks, 25 weeks, and 23 weeks.

The cost-minimizing procedure begins by first crashing the critical activities that have the least unit additional cost, and continues by crashing activities with increasingly larger unit additional costs. To reduce the project duration from 29 weeks to 27 weeks, the best activity to crash is *Design mechanical equipment*, which can be reduced at an additional cost of $10,000 per week. Total added costs for shaving two weeks from the schedule is $20,000.

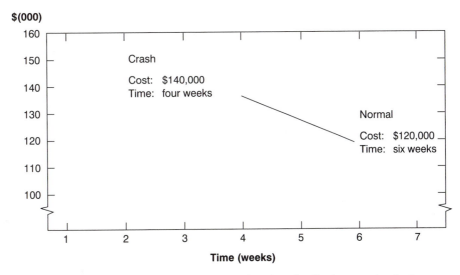

FIGURE 5.10   Time/cost curve for accelerating the *Design mechanical equipment* activity

The crashing arrangement to reduce the project duration to 25 weeks becomes more involved:

1. If *Design electrical equipment* is reduced one week at an additional cost of $15,000, it will be necessary to reduce *Fabricate mechanical equipment* one week at an additional cost of $8,000 for a total cost of $23,000.
2. For the second-week reduction, the best arrangement is to speed up by one week the concurrent activities, *Construct phase I* and *Install mechanical and electrical equipment,* each of which will add $25,000 to the cost. The total additional cost to reduce the project duration to 25 weeks is $73,000.
3. To reduce project duration to 23 weeks requires examining the duration of all of the project paths and determining which activities need to be reduced to shorten the project duration from 25 weeks to 23 weeks.
4. Using the Christopher Design/Build Project planning diagram as a worksheet to examine each path for a maximum of 23 weeks, we see that five activities must each be shortened one week to enable a two-week reduction:
   - *Fabricate and erect structural steel* for an added cost of $15,000;
   - *Construct phase I building,* $20,000;
   - *Design building,* $30,000;
   - *Install mechanical & electrical equipment,* $25,000;
   - *Construct phase II building,* $25,000.
   Total added costs for cutting the schedule two additional weeks: $110,000.

Figure 5.12(a) shows the tabulation of the effect of the added costs to the project when decreasing the project duration, and Figure 5.12(b) displays the results of each project path as activities are reduced to reach the desired project durations.

i-j	Project Activity Description	Normal Duration (Weeks)	Normal Cost	Crash Duration (Weeks)	Crash Cost	Unit Cost (per Week)
1-2	Design structural steel	2	$ 40,000.00	2	$ 40,000.00	$ 0.00
1-3	Design building	13	420,000.00	9	540,000.00	30,000.00
1-4	*Design mechanical equipment	6	120,000.00	4	140,000.00	10,000.00
2-3	Fabricate and erect structural steel	10	220,000.00	7	265,000.00	15,000.00
3-7	Construct phase I building	6	420,000.00	5	440,000.00	20,000.00
4-5	*Design electrical equipment	4	80,000.00	3	95,000.00	15,000.00
4-6	Fabricate mechanical equipment	9	90,000.00	6	114,000.00	8,000.00
4-8	Design interior items	3	40,000.00	1	55,000.00	15,000.00
5-6	*Procure electrical equipment	7	40,000.00	5	60,000.00	10,000.00
6-7	*Install mechanical and electrical equipment	6	200,000.00	4	250,000.00	25,000.00
7-11	*Construct phase II building	6	450,000.00	4	500,000.00	29,000.00
8-9	Procure interior items	4	40,000.00	3	50,000.00	10,000.00
9-10	Install interior items	10	240,000.00	6	320,000.00	20,000.00
	Total Cost		2,400,000.00			

*Critical activities.

FIGURE 5.11　Normal/crash data for the Christopher Design/Build Project

(a)

Project Duration (Weeks)	Normal Cost	Project Activity Node No./Description	Reduced Time (Weeks)	Crash Cost	Normal and Crash Costs
29	$2,400,000				
27		1-4 Design mechanical equipment	2	$20,000	$2,420,000
25		4-5 Design electrical equipment	1	$15,000	
		4-6 Fabricate mechanical equipment	1	8,000	
				23,000	
		7-11 Construct phase II building	1	25,000	
		6-7 Install mechanical and electrical equipment	1	25,000	2,493,000
				73,000	
23		2-3 Fabricate and erect structural steel	1	15,000	
		3-7 Construct phase I building	1	20,000	
		1-3 Design building	1	30,000	
		6-7 Install mechanical and electrical equipment	1	25,000	
		7-11 Construct phase II building	1	25,000	2,603,000
				110,000	

(b)

Project Paths	Durations			
1 - 4 - 5 - 6 - 7 - 11	29	27	25	23
1 - 2 - 3 - 7 - 11	29	27	25	23
1 - 3 - 7 - 11	24	24	24	23
1 - 4 - 8 - 9 - 10	25	25	25	21
	22	22	22	22

FIGURE 5.12 (a) Calculating crash costs for the Christopher Design/Build Project; (b) Christopher Design/Build Project paths

As mentioned earlier, the costs associated with project activities consist of direct (sometimes called variable) costs, including labor, material, and equipment. Other costs identified with the project include such items as office expenses, insurance, interest on loans to finance the project, supervision, general office personnel, and office space. Commonly referred to as indirect costs, these can amount to an appreciable sum. (The complete project funds package allows funds for the indirect costs that are not included with the direct costs.)

Snyder Engineering has assessed $580,000 in indirect costs for the 29-week Christopher Design/Build Project. Snyder has also determined that the total indirect costs would decrease by $20,000 per week as the project is nearing completion.

A summary of the indirect costs and their effect on total costs when crashing the project is as follows:

Project Duration (Weeks)	Normal and Crash Costs	Indirect Costs	Total Direct and Indirect Costs
29	$2,400,000	$580,000	$2,980,000
27	2,420,000	540,000	2,860,000
25	2,493,000	500,000	2,993,000
23	2,603,000	460,000	3,063,000

Note the impact of indirect costs on the project. It is an item that bears consideration when determining the costs of a project. It is interesting to note that reducing the project duration from 29 weeks to 27 weeks actually reduces total project costs by $120,000. The reduction is the result of eliminating all of the indirect costs—no more insurance or loans to finance the job are needed, the office space can be vacated, office supplies and expenses are gone, and supervision and office personnel leave for another project.

Further reduction of the project duration to 25 weeks will increase costs by just $13,000 or 0.4 percent over the total project's direct and indirect costs. Reducing the project duration to 23 weeks would be a $103,000 (3.5 percent over) increase in total cost. Figure 5.13 illustrates the complete effect of both direct and indirect costs on the Christopher Design/Build Project.

For decision-making purposes, this cost-minimizing exercise is prepared early in the project. These options are connected with the personnel/labor planning, and if they are adopted, they result in a new time and cost schedule. The cost-minimizing value will also be recognized in the event the project begins to experience serious delays and requires crashing specific work activities to get back on schedule.

Project Duration (Weeks)	Direct Costs	Indirect Costs			Total Costs Direct and Indirect
		Normal	Crash	Total	
29	$2,400,000.00	$580,000.00	$ 0.00	$580,000.00	$2,980,000.00
27	2,420,000.00		(40,000.00)	540,000.00	2,860,000.00
25	2,493,000.00		(40,000.00)	500,000.00	2,993,000.00
23	2,603,000.00		(40,000.00)	460,000.00	3,063,000.00

(Indirect costs decrease at the rate of $20,000 per week. Indirect costs include office expenses, insurance, interest on loans to finance the project, supervision, general office personnel, and temporary office space.)

FIGURE 5.13 Calculations of indirect costs and total costs to crash the Christopher Design/Build Project

USING THE COMPUTER

Computer-generated cost schedule reports can be prepared concurrently with the project timing schedules. This is a decided advantage at the beginning of a project, when the project team can compare these reports with the objectives. With a computer, the project team can revise and correct quickly and accurately. Knowing an accurate distribution of costs can be computer generated is important for budget planning.

Using a worksheet before inputting the cost data in the computer collects it in an organized fashion that will save time by reducing the number of corrections as data is entered. (See Figure 5.14.) For the computer program being used in this project, the cost levels are expressed on a daily basis. Before producing reports, use selected symbols—TC (daily costs) and $C (accumulated costs)—and include them in the abbreviations file.

1. Select option **4**: Do Some Housekeeping?
2. Select option **2**: Work with Abbreviations Dictionary.
3. Select option **2**: Change or Display an Abbreviations File.
4. Select option **1**: Change or Add Abbreviations.
5. Type the abbreviations (or symbols) and descriptions.
   a. TC DAILY COSTS ($)
   b. $C ACCUMULATED COSTS ($)
6. Select option **5**: Save This Version and Return (to Main Menu).

Next, access the screen to enter the cost data through the main menu:

1. Select option **1**: Produce an Up-to-Date Plan?
2. Do You Want to Work With: Press **DBP version 2** (or version that you are using).

3. Press **F2** to enter the resources data screen.
4. Type **1** at *Command?*
5. On Line 1: Type (for the corresponding node numbers) **TC2857 $C2857** [ENTER].

```
COMMAND?                           Current Plan: DBP      version 5

LINE    FROM    TO      RESOURCES ALLOCATED PER UNIT TIME        ?=HELP

 1 :     1   -> 2              TC2857 $C2857
 2 :         -> 3              TC4615 $C4615
 3 :         -> 4              TC2857 $C2857
 4 :     2   -> 3              TC3143 $C3143
 5 :     3   -> 7              TC10000  $C10000
 6 :     4   -> 5              TC2857 $C2857
 7 :         -> 6              TC1429 $C1429
 8 :         -> 8              TC2857 $C2857
 9 :     5   -> 6              TC816 $C816
10:      6   -> 7              TC4762 $C4762
11:      7   -> 11            TC10417  $C10417
12:      8   -> 9             TC1429 $C1429
13:      9   -> 10            TC3429  $C3429




    1DESCRP 2RESRCE 3INSERT 4NEW PG 5HDINGS 6GO TOP 7ZOOM   8BARCHT 9ESCAPE 10HELP
```

6. Press [ENTER], then [ESC] when all lines are completed.
7. TYPE **2** at *Enter a New Time Now Day Number*, then [ENTER].

You are now ready to print cost reports. For the daily cost report:

1. Press option **5**: Histogram for a Resource [ENTER].
2. Press **1**: That All Activities Start As Early As Possible [ENTER].
3. Type **2** for The Day Number That You Want This Report to Start From.
4. For the Resource to be Analyzed in this Report: Type **TC** [ENTER].
5. Select **8**: Start or Cancel the Run.
6. Press [ENTER] to start the run.

Figure 5.15 shows the cost histogram (or cost distribution graph) for the Christopher Design/Build Project. It displays the daily costs when using the early start schedule. While this graph does not display exact cost values, it is close enough for some early cost analysis. From this graph Snyder Engineering, satisfied with the distribution of costs for at least the first half of the project, gives tentative approval of the budget.

Costs Worksheet

Project Name:
Design/Build Project

Start Node	End Node	Duration (Days)	Description	Costs In Dollars		
				Total Costs	Daily Costs	
					TC	$C
Start	1	—	————	— —	—	—
1	2	14	Design structural steel	40,000	2,857	2,857
1	3	91	Design building	420,000	4,615	4,615
1	4	42	Design mechanical equipment	120,000	2,857	2,857
2	3	70	Fabricate and erect structural steel	220,000	3,143	3,143
3	7	42	Construct phase I building	420,000	10,000	10,000
4	5	28	Design electrical equipment	80,000	2,857	2,857
4	6	63	Fabricate mechanical equipment	90,000	1,429	1,429
4	8	14	Design interior items	40,000	2,857	2,857
5	6	49	Procure electrical equipment	40,000	816	816
6	7	42	Install mechanical and electrical equipment	200,000	4,762	4,762
7	11	42	Construct phase II building	450,000	10,714	10,714
8	9	28	Procure interior items	40,000	1,429	1,429
9	10	70	Install interior items	240,000	3,429	3,429
10	11	—	Dummy	— —	—	—
11	F	—	————			

TC—Project Activity Costs
$C—Project Costs

FIGURE 5.14 Costs worksheet

During the first half of the project, the distribution graph helps Snyder Engineering set up payments from the owner to compensate for at least all of the costs to be incurred. (If Snyder did not, the contractor might be forced to borrow money, paying interest that usually is not included in a proposal.) From this report, Snyder Engineering will develop a schedule of payments (for the owner's approval) for the work to be performed. To validate the payment schedule Snyder Engineering must stay on schedule. Using the early start schedule will give it the advantage of using the available float. The late start schedule does not allow for any glitches in the schedule—all project activities would be on critical paths and any type of delay, in any activity, would adversely affect the completion date.

Accumulated Cost Report

The accumulated cost chart is a computer-generated graphic picture of the buildup of authorized costs at specific times during the project. It is a plotted cost

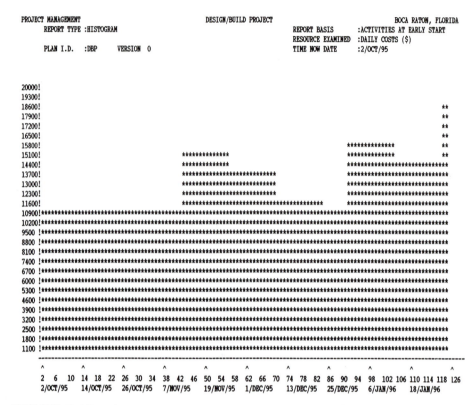

```
PROJECT MANAGEMENT                          DESIGN/BUILD PROJECT                              BOCA RATON, FLORIDA
        REPORT TYPE :HISTOGRAM                                            REPORT BASIS      :ACTIVITIES AT EARLY START
                                                                         RESOURCE EXAMINED :DAILY COSTS ($)
        PLAN I.D.   :DBP     VERSION  0                                  TIME NOW DATE     :2/OCT/95
```

FIGURE 5.15   Cost histogram for the Christopher Design/Build Project

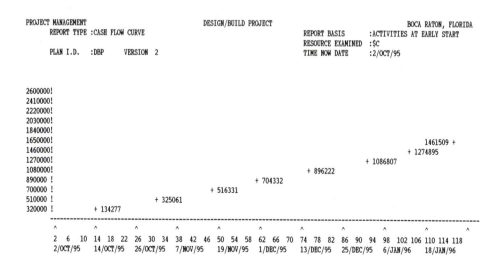

```
PROJECT MANAGEMENT                          DESIGN/BUILD PROJECT                              BOCA RATON, FLORIDA
        REPORT TYPE :CASH FLOW CURVE                                     REPORT BASIS      :ACTIVITIES AT EARLY START
                                                                         RESOURCE EXAMINED :$C
        PLAN I.D.   :DBP     VERSION  2                                  TIME NOW DATE     :2/OCT/95
```

FIGURE 5.16   Accumulated cost curve for the Christopher Design/Build Project,
October 5, 1995 through January 30, 1996

curve, also known as the cash flow curve. The cash flow curve utilizes the same data used to develop the cost schedule and shows the accumulated costs at any given date in the project.

To produce the cash flow curve, use the same instructions for obtaining a cost histogram (see last section) until you reach step 4:

4. For the Resource to be Analyzed in this Report: Type **$C!** [ENTER].

Figure 5.16 is the accumulated cost curve for one period of the Christopher Design/Build Project. The curve shows about $1,500,000 (or 58 percent of the total costs) will be incurred during this 17-week period, about 65 percent of the total project duration. This schedule suggests a good distribution of costs over the life of the project, and Snyder Engineering is satisfied to proceed with the start of the project without any changes at this time.

SUMMARY

Managing costs is one of the most essential parts of project management. Effective management of costs and time is the very essence of project management. While time delays may be tolerated to some degree, funds allocated to many projects cannot be exceeded. Some time delays may require additional funds that are literally impossible to obtain.

There is an analogy between managing time and managing costs. Just as successful projects have disciplined methods to plan, schedule, and control the project timing, project costs may need to be contained using similar methods:

- **Planning** uses the planned budget to develop the authorized cost of each project activity. Plans are prepared by personnel with experience and knowledge, who will be responsible for the integrity of the specific project activities.
- **Scheduling** distributes the costs over the life of the project. In addition to the cost tabulation format, the bar chart cost schedule provides a graphic display of the total costs for each project activity.
- In **cost control**, once a project is under way, actual expenditures are compared periodically with the planned (budgeted) costs.

The indicated cost outcome method is an early warning system to ensure containment of project spending within the budget. The indicated cost outcome report reviews and evaluates the spending status of the project—what has been spent and what is planned to be spent. The report is designed to warn project personnel about evidence of excessive spending, allowing enough time to correct it. An item that indicates an overspending (overrun) trend by the end of the project will be targeted early enough in the course of the project to allow remedial action to be taken.

Cost minimizing is a project management technique that allows reduction of the project duration at the least added expense. To set this method in motion requires completion of the project planning diagram, the time schedule, and the cost schedule. Critical items are investigated for possible crashing through additional personnel, overtime for existing personnel, additional equipment or resources, or combinations of these. The method allows for a "cut and try" procedure that determines what combination of project activities will yield an earlier date for completion with the least additional costs.

The cost-minimizing method introduces the indirect costs associated with the project. Indirect costs include office expenses, insurance, interest on loans to finance the project, supervision, and temporary office space. The effects of indirect costs are significant on the total costs of the project—on certain occasions, reducing the project duration could result in reduction of the actual total cost of the project because of the indirect costs.

Preparing this type of information during the early stages of the project will be useful, as it will be available as part of the analysis if schedule adjustments are being considered to resolve potential serious delays.

Exercises

1. What are the two most essential functions in project management?
2. For this exercise use the planning diagram in Figure 1.2; the time estimates for each activity shown in Exercise 3, Chapter 3; and the following cost information:

Activity	Weeks	Total Cost
Structural engineering	3	$ 24,000
Fabricate and erect structural steel	10	150,000
Construction engineering	5	40,000
Phase I construction	12	350,000
Phase II construction	20	400,000
Award contract	4	8,000
Equipment engineering	8	60,000
Deliver and fabricate equipment	12	600,000
Phase I installation	10	200,000
Phase II installation	8	150,000
Long-lead items—engineer	6	50,000
Long-lead items—procure	18	400,000

   a. Construct a bar chart cost schedule. (Suggestion: Start with the bar chart time schedule developed for this project in the Chapter 3 exercises.)
   b. Develop a tabulated cost schedule in 15-week increments. (Use the early start schedule for this tabulation.)
   c. Construct a cost distribution graph in 15-week increments.

3. Using the data in Exercise 2:
   a. Prepare a computer-generated cost histogram for the total project.
   b. Prepare a computer-generated accumulated cost curve for the complete project.
4. Referring to the Christopher Design/Build Project, Snyder Engineering wants to know what costs will be incurred from January 30, 1996 to the end of the project if they maintain the early start schedule.
   a. Prepare a computer-generated cost histogram for the January 1, 1996 through April 21, 1996 period.
   b. Prepare an accumulated cost curve for the above period.

Personnel/Labor Planning

THIS CHAPTER WILL COVER

- Procedure for planning resources
- Comparing supply with demand
- Planning resources with the computer

Planning, scheduling, and controlling projects as presented so far have had no restrictions on labor. While some projects may have sufficient resources to deal with their requirements, many situations will require careful planning of available resources to meet the demand. For efficiency, resources also need to be distributed evenly over the span of the project.

Consistent employment has a number of benefits. When employees feel their jobs are permanent, their morale is higher. In addition, continually laying employees off, then rehiring them, is costly and inefficient. Companies that earn a reputation for such practices may have difficulty finding workers to meet peak demands. Another intangible inefficiency is possible: When employees detect that their work is nearing an end, they may begin a subtle slowdown.

Many firms experience heavy and light periods of employment demand. Although these situations exist more in construction projects and architectural and engineering fields, many other ventures experience inconsistent demands. Specific resource planning procedures are available, however, that can assist in leveling out the demands of personnel and labor. Used with the planning and scheduling techniques previously discussed, they are very helpful.

Planning the labor and personnel requirements for many projects is as necessary as planning timing and costs. Inconsistent resource demands are not unusual. Detecting and correcting these situations early in the project will eliminate or at least minimize any additional costs that could be incurred. Specific techniques to be discussed in this chapter use the early warning approach to help stabilize labor and personnel demands. Later in the chapter, the computer will be used to enhance the effectiveness of these procedures in certain situations.

RESOURCE PLANNING PROCEDURES

Planning resources is essentially an effort to match the resource demands of a project with the availability (or supply). The leveling process is actually a form of resource scheduling; as such, it is related to timing of the project. Activities are shifted within their earliest/latest range to note changes in demand. The technique follows this sequence:

- First, examine activities with the most float values for rescheduling.
- Next, use activities with relatively small float values.
- Finally, if the need to continue the leveling process is critical, use the activities that have no float.

An important rule is to do as much adjusting as necessary for leveling purposes without using the critical activities in the project. One of the objectives of this process is not to affect the project duration.

To start the resource leveling method, complete these items: (1) project objectives, (2) project planning diagram, (3) project schedule (including the bar chart time schedule), and (4) preparing the total float tabulation (identifying the critical and noncritical activities). These steps follow:

1. Place the daily resources needed to complete each activity on the bottom of each bar associated with that activity.
2. Divide the bar chart into daily increments and total up the daily resources, showing the tabulation of each specific resource at the bottom of the bar chart. See Figure 6.1 for a partial tabulation (weeks 1–4 and 25–29) of the Christopher Design/Build Project.
3. This tabulation will show the amount of each resource expended for each period. Consider adjusting the scheduling of the activities if there is a wide variation in the daily resources expended and if demand is inconsistent over the span of the project.
4. The first scheduling adjustments for leveling purposes should affect individual activities: Manually move the activity with the highest float value in the peak location, total the demand after each adjustment, and note the leveling effect to determine if further leveling is necessary.
5. If all the scheduling adjustments are still unsatisfactory, there are three alternatives:
   a. Authorize overtime for the critical personnel in the critical periods.
   b. If overtime is not an acceptable option, then attempt to add more personnel.
   c. If the above two alternatives are not practical, then steps need to be taken to extend the project duration.

RESOURCE PLANNING: CHRISTOPHER DESIGN/BUILD PROJECT

To illustrate the manual method for personnel scheduling and the leveling procedure, we will use the Christopher Design/Build Project. We have set these objectives:

- The project duration must remain at 29 weeks.
- The number of personnel allocated for this project includes 8 engineers, 15 designers, 6 architects, and 30 skilled personnel.
- A reasonable constant crew size should be maintained within the allowable limits, minimizing peaks and valleys of demand.
- Overtime should be kept at a minimum.

As a basis for planning, a number of previously developed documents are needed:

- The project planning diagram (Figure 6.2) should show the personnel required to complete each activity. Place under the appropriate activity arrows the daily requirements of engineers, designers, architects, and skilled personnel.
- Use the total float calculations developed in Chapter 3 to highlight the critical and noncritical activities (shown in Figure 6.3). Investigating one or two activities with the largest float at the onset for scheduling adjustments may be all that is needed to resolve any leveling problems.

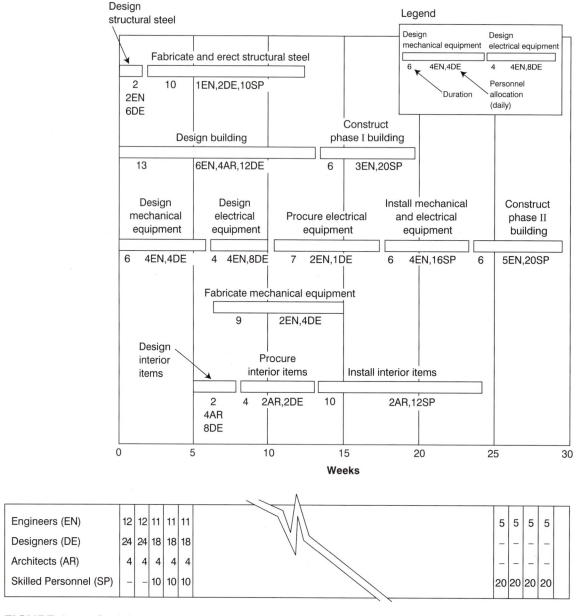

FIGURE 6.1   Partial tabulation of personnel distribution for the Christopher Design/Build Project

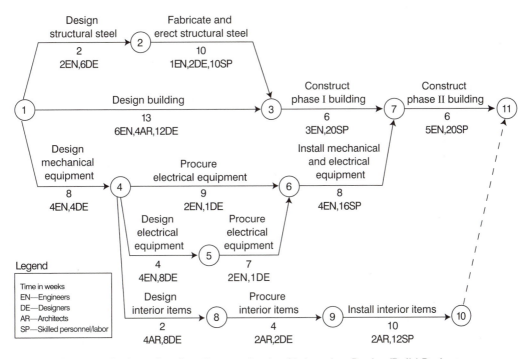

FIGURE 6.2   Project planning diagram for the Christopher Design/Build Project

- To aid in beginning the personnel leveling operation, place the daily personnel requirements for each activity under its respective bar on the bar chart time schedule as shown in Figure 6.4(a). Start with the early start schedule, the preferable schedule for setting a project in motion.
- Total for each week, as shown in the tabular format depicted in Figure 6.4(b), the average daily number of engineers, designers, architects, and skilled personnel. It is the first examination of the personnel loadings (total number of personnel needed daily) for the project reflecting the early start schedule. This initial loadings analysis reveals that the personnel at various periods during the project are appreciably higher than the allowable supply for this project. Most significant are the designers and skilled personnel.
- The required number of designers exceeds the authorized supply (15) during 13 of the 17 weeks. The heaviest demand occurs during the first half of the Christopher Design/Build Project. The available 15 designers do not match up to a peak demand of 34 for two weeks, 28 for two weeks, and 24 for another two weeks. These are concerns that need analysis and need resolution.
- Also a concern is the need for available skilled personnel, which peaks at 48 for two weeks. At other periods the supply can handle the demand, but demand is questionable for various periods during the length of the project.

Activity (Node Numbers)	Description	Latest Finish Time	(−)	Earliest Start Time	(−)	Time Estimate for Completing Activity	(=)	Total Float
1-4	Design mechanical equipment	6		0		6		0*
4-5	Design electrical equipment	10		6		4		0*
5-6	Procure electrical equipment	17		10		7		0*
6-7	Install mechanical and electrical equipment	23		17		6		0*
7-11	Construct phase II building	29		23		6		0*
1-2	Design structural steel	7		0		2		5
1-3	Design building	17		0		13		4
2-3	Fabricate and erect structural steel	17		2		10		5
3-7	Construct phase I building	23		13		6		4
4-6	Procure electrical equipment	17		6		9		2
4-8	Design interior items	15		6		2		7
8-9	Procure interior items	19		8		4		7
9-10	Install interior items	29		12		10		7

*Critical path.

FIGURE 6.3   Total float tabulation for the Christopher Design/Build Project

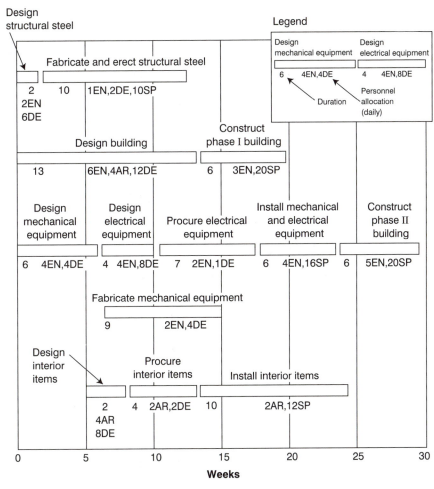

(a)

Design structural steel

Legend
Design mechanical equipment | Design electrical equipment
6   4EN,4DE | 4   4EN,8DE
Duration   Personnel allocation (daily)

Fabricate and erect structural steel
2 2EN 6DE   10   1EN,2DE,10SP

Design building
13   6EN,4AR,12DE

Construct phase I building
6   3EN,20SP

Design mechanical equipment
6   4EN,4DE

Design electrical equipment
4   4EN,8DE

Procure electrical equipment
7   2EN,1DE

Install mechanical and electrical equipment
6   4EN,16SP

Construct phase II building
6   5EN,20SP

Fabricate mechanical equipment
9   2EN,4DE

Design interior items
2 4AR 8DE

Procure interior items
4   2AR,2DE

Install interior items
10   2AR,12SP

Weeks 0   5   10   15   20   25   30

(b)

Engineers (EN)	12	12	11	11	11	11	13	13	13	13	11	11	10	7	7	5	5	7	7	4	4	4	4	5	5	5	5	5	5
Designers (DE)	22	22	18	18	18	18	34	34	28	28	21	21	17	5	5	1	1												
Architects (AR)	4	4	4	4	4	4	8	8	6	6	6	6	6	2	2	2	2	2	2	2	2	2							
Skilled Personnel (SP)			10	10	10	10	10	10	10	10	10	12	32	32	32	32	48	48	28	28	28	16	20	20	20	20	20	20	20

FIGURE 6.4   (a) Bar chart time schedule (early start) with personnel allocation; (b) weekly average daily personnel tabulation

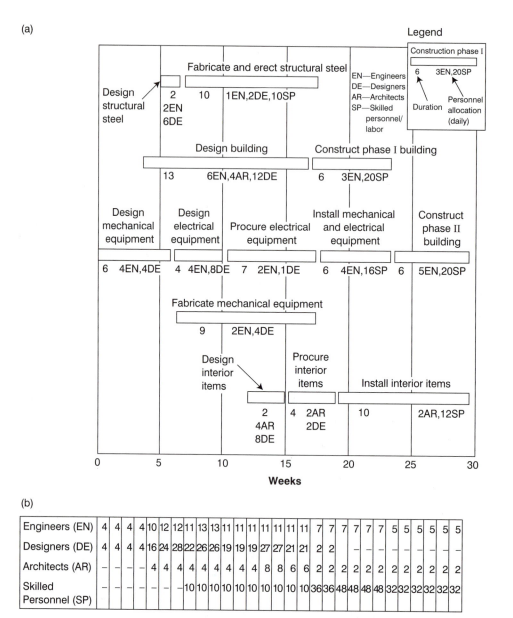

(a)

(b)

Engineers (EN)	4	4	4	4	10	12	12	11	13	13	11	11	11	11	11	11	11	7	7	7	7	7	7	5	5	5	5	5	5
Designers (DE)	4	4	4	4	16	24	28	22	26	26	19	19	19	27	27	21	21	2	2	—	—	—	—	—	—	—	—	—	—
Architects (AR)	—	—	—	—	4	4	4	4	4	4	4	4	4	8	8	6	6	2	2	2	2	2	2	2	2	2	2	2	2
Skilled Personnel (SP)	—	—	—	—	—	—	—	10	10	10	10	10	10	10	10	10	10	36	36	48	48	48	48	32	32	32	32	32	32

FIGURE 6.5 (a) Bar chart time schedule (late start) with personal allocation; (b) weekly average daily personnel tabulation

The irregular demand for both designers and skilled personnel over extended periods suggests the impracticability of adjusting schedules of individual activities. It would be more desirable to investigate the resource situation when all activities are scheduled at their latest start times.

To check the requirements at the latest start schedule, use the same procedure as outlined above. The completed check begins with the bar chart in Figure 6.5(a), which shows the average daily personnel requirements for each activity scheduled at its latest start date. Correspondingly, Figure 6.5(b) shows the tabular format of each week's average daily requirements of engineers, designers, architects, and skilled personnel when the project is scheduled at the latest start. This schedule appears to have a more feasible personnel allocation than the early start schedule:

- Loadings when using the latest start schedule are an improvement over the earliest start schedule.
- Designer demand is still high, but to a lesser degree. The number of designers required during peak periods drops from 34 when using the early start schedule to 28 when using the late start schedule.
- Skilled personnel during peak periods are the same with both schedules. However, the late start schedule allows for a more consistent demand compared to the erratic demands of the early start schedule.

While there already appears to be enough support for the late start schedule, the analysis needs to continue for further assurance. Another check is to compare the additional personnel hours (and thus additional costs) that either schedule may incur. A graphic representation of the personnel loadings will clearly show this comparison.

Graphic Loadings

Graphing the daily staff requirements will readily show the periods of heavy and light demand. Constructing a graphic load chart is not difficult, yet it is most valuable in clarifying the personnel picture. The horizontal scale represents the weeks (or whatever unit is to be used), and the vertical scale represents the number of designers or skilled personnel (the critical personnel in the Christopher Design/Build Project) needed.

The graphic load chart for the designers (see Figure 6.6) shows the supply and demand for both the early start and late start schedules. This chart showing the two schedules can be useful for comparing the amount of additional hours required. The area between the supply line and the upper limits of the requirements line represents the total additional hours.

FIGURE 6.6 Graphic load chart of designer allocation for the Christopher Design/Build Project

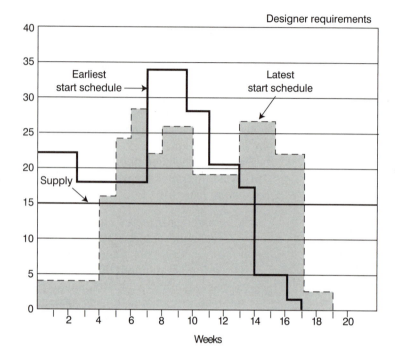

The additional hours incurred using the earliest start time are calculated and then tabulated as follows:

Weeks	Designers Demand	Designers Available	Additional Designers Required	Additional Hours Required
1-2	24	15	9	1,008
3-6	18	15	3	672
7-8	34	15	19	2,128
9-10	28	15	13	1,456
11-12	21	15	6	672
13	17	15	2	112
				6,048

When the late start schedule is used, the additional hours are tabulated as follows:

Weeks	Designers Demand	Available	Additional Designers Required	Additional Hours Required
5	16	15	1	56
6	24	15	9	504
7	28	15	13	728
8	22	15	7	392
9-10	26	15	11	1,232
11-13	19	15	4	672
14-15	27	15	12	1,344
16-17	21	15	6	672
				4,600

The results of these tabulations appear to show a convincing case that the latest start schedule should be used in the implementation of the Christopher Design/Build Project. First, 1,224 more designer hours are required using the early start schedule—an impressive burden to carry. In addition, most of these additional hours occur in the first 10 weeks of the project. Paying for this additional time might pose a cash flow problem. Performance payments to the contractor may trail completed work by a month or two.

Project timing still can provide a rationale for considering the early start schedule, however. With the early start schedule, the project manager has the availability of float on many of the items, in contrast to the late start schedule, in which all project activities would be critical. Options on what course to follow are presented to those responsible for this project. The decisions may be subjective, but supported with information using sound project management procedures. With computer support enhancing the information supplied, the decision makers will help ensure a satisfactory solution.

RESOURCE PLANNING BY COMPUTER

Once you understand the reasoning behind resource leveling methods, using project management software becomes more useful. The computer can handle the necessary calculations for resource planning, generating information so that decisions are more accurate and much faster.

The computer employs the same procedure used in developing the manual calculations. To illustrate the computer application, we will again use the Christopher Design/Build Project as the sample project.

The Christopher Design/Build Project planning diagram, with the labor/personnel allocated for each activity (refer to Figure 6.2), will be used as the source document for the input data. Supplementing the planning diagram are the computerized schedule and bar chart reports shown in Figures 6.7(a) and 6.7(b), respectively. (The Chapter 3 computer discussions include instructions on producing these printouts.)

Using a worksheet organizes the labor/personnel data to ease the computer input process (see Figure 6.8). This approach will save time as it helps detect discrepancies and errors early. Changes that occur when inputting directly into the computer can take important time away from receiving the necessary reports to begin the analysis.

(a)

```
PROJECT MANAGEMENT                         DESIGN/BUILD PROJECT                        BOCA RATON, FLORIDA
    REPORT TYPE :STANDARD LISTING                               PRINTING SEQUENCE :Earliest Activities First
                                                                SELECTION CRITERIA :ALL
    PLAN I.D.   :DBP     VERSION 1                              TIME NOW DATE      : 2/OCT/95
==========================================================================================================
    ACTIVITY DESCRIPTION            EARLIEST      EARLIEST     LATEST       LATEST      DURATION FLOAT
                                    START         FINISH       START        FINISH
==========================================================================================================
    1-   4 DESIGN MECHANICAL EQUIPMENT    2/OCT/95     12/NOV/95     2/OCT/95     12/NOV/95      42      0 *
    1-   3 DESIGN BUILDING               2/OCT/95     31/DEC/95    30/OCT/95     28/JAN/96      91     28
    1-   2 DESIGN STRUCTURAL STEEL       2/OCT/95     15/OCT/95     6/NOV/95     19/NOV/95      14     35
    2-   3 FABRICATE & ERECT STRUCTURAL STEEL 16/OCT/95 24/DEC/95 20/NOV/95    28/JAN/96      70     35
----------------------------------------------------------------------------------------------------------
    4-   5 DESIGN ELECTRICAL EQUIPMENT  13/NOV/95     10/DEC/95    13/NOV/95     10/DEC/95      28      0 *
    4-   6 FABRICATE MECHANICAL EQUIPMENT 13/NOV/95   14/JAN/96    27/NOV/95     28/JAN/96      63     14
    4-   8 DESIGN INTERIOR ITEMS        13/NOV/95     26/NOV/95     1/JAN/96     14/JAN/96      14     49
    8-   9 PROCURE INTERIOR ITEMS       27/NOV/95     24/DEC/95    15/JAN/96     11/FEB/96      28     49
----------------------------------------------------------------------------------------------------------
    5-   6 PROCURE ELECTRICAL EQUIPMENT 11/DEC/95     28/JAN/96    11/DEC/95     28/JAN/96      49      0 *
    9-  10 INSTALL INTERIOR ITEMS       25/DEC/95      3/MAR/96    12/FEB/96     21/APR/96      70     49
    3-   7 CONSTRUCT PHASE I BUILDING    1/JAN/96     11/FEB/96    29/JAN/96     10/MAR/96      42     28
    6-   7 INSTALL MECH & ELECT EQUIP   29/JAN/96     10/MAR/96    29/JAN/96     10/MAR/96      42      0 *
----------------------------------------------------------------------------------------------------------
    7-  11 CONSTRUCT PHASE II BUILDING  11/MAR/96     21/APR/96    11/MAR/96     21/APR/96      42      0 *
==========================================================================================================
```

(b)

```
PROJECT MANAGEMENT                         DESIGN/BUILD PROJECT                        BOCA RATON, FLORIDA
    REPORT TYPE :COMPRESSED PERIOD BARCHART                       PRINTING SEQUENCE :Earliest Activities First
                                                                  SELECTION CRITERIA :ALL
    PLAN I.D.   :DBP     VERSION 1                                TIME NOW DATE      : 2/OCT/95
=====================================1995=====================1996=========================================
    PERIOD COMMENCING DATE          !2   !27  !21  !16  !10  !4   !29  !25  !19  !14  !8   !3   !
    MONTH                           !OCT !    !NOV !DEC !JAN !FEB !    !MAR !APR !MAY !JUN !JUL !
    PERIOD COMMENCING TIME UNIT     !2   !27  !52  !77  !102 !127 !152 !177 !202 !227 !252 !277 !
==========================================================================================================
    1-   4 DESIGN MECHANICAL EQUIPMENT  !CCCCCC!CCCCC !    !    !    !    !    !    !    !    !    !    !
    1-   3 DESIGN BUILDING              !======!======!======!======!..!......!    !    !    !    !    !    !
    1-   2 DESIGN STRUCTURAL STEEL      !====..!.......!.    !    !    !    !    !    !    !    !    !    !
    2-   3 FABRICATE & ERECT STRUCTURAL S !    ===!======!======!===....!......!    !    !    !    !    !    !
----------------------------------------------------------------------------------------------------------
    4-   5 DESIGN ELECTRICAL EQUIPMENT  !    !    CC!CCCCCC! !    !    !    !    !    !    !    !    !
    4-   6 FABRICATE MECHANICAL EQUIPMENT !   !    ==!======!=======!==... !    !    !    !    !    !    !    !
    4-   8 DESIGN INTERIOR ITEMS        !    !    ==!==....!......!...    !    !    !    !    !    !    !    !
    8-   9 PROCURE INTERIOR ITEMS       !    !    !    ====!=====..!......!...    !    !    !    !    !    !
----------------------------------------------------------------------------------------------------------
    5-   6 PROCURE ELECTRICAL EQUIPMENT !    !    !    C!CCCCCCC!CCCCC !    !    !    !    !    !    !
    9-  10 INSTALL INTERIOR ITEMS       !    !    !    !    ====!=====!======!==....!........!.    !    !    !
    3-   7 CONSTRUCT PHASE I BUILDING   !    !    !    !    ===!======!===...!....    !    !    !    !    !
    6-   7 INSTALL MECH & ELECT EQUIP   !    !    !    !    CC!CCCCCCC!CCCC !    !    !    !    !    !
----------------------------------------------------------------------------------------------------------
    7-  11 CONSTRUCT PHASE II BUILDING  !    !    !    !    !    !    CCC!CCCCCCCC!C !    !    !    !
==========================================================================================================
Barchart Key:-  CCC :Critical Activities   === :Non Critical Activities   NNN :Activity with neg float   ... :Float
```

FIGURE 6.7 (a) Computer-generated schedule; (b) computer-generated bar chart report for the Christopher Design/Build Project

Activities/Resources Worksheet

Project Name:
Christopher Design/Build Project

Start Node	End Node	Duration (Days)	Description	Resources (Daily Requirements)			
				EN	AR	DE	SP
Start	1	—	————	—	—	—	—
1	2	14	Design structural steel	2	—	6	—
1	3	91	Design building	6	4	12	—
1	4	42	Design mechanical equipment	4	—	4	—
2	3	70	Fabricate and erect structural steel	1	—	2	10
3	7	42	Construct phase I building	3	—	—	20
4	5	28	Design electrical equipment	4	—	8	—
4	6	63	Fabricate mechanical equipment	2	—	4	—
4	8	14	Design interior items	—	4	8	—
5	6	49	Procure electrical equipment	2	—	1	—
6	7	42	Install mechanical and electrical equipment	4	—	—	16
7	11	42	Construct phase II building	5	—	—	20
8	9	28	Procure interior items	—	2	2	—
9	10	70	Install interior items	—	2	—	12
10	11	—	Dummy	—	—	—	—
11	F	—	————				
1	Milestone		Start project	EN—Engineers			
3	Milestone		Complete structural steel	AR—Architects			
6	Milestone		Complete electrical equipment–	DE—Designers			
			design and procure	SP—Skilled Personnel/Labor			
7	Milestone		Complete mechanical and electrical;				
			Complete phase I				
10	Milestone		Complete architectural				
11	Milestone		Complete project				

FIGURE 6.8   Activities/resources worksheet for labor/personnel data

Personnel/labor levels are usually reported on a daily basis. The selected symbols used for this project that are to be entered as basic computer data are **EN** (engineers), **AR** (architects), **DE** (designers), and **SP** (skilled personnel/labor). Germane to using this computer program is the completion of the abbreviations file before producing any reports. Start with the main menu to open the abbreviations file and to reach the existing DBPA file:

1. Press **4**: Do Some Housekeeping?
2. Press **2**: Work with Abbreviations Dictionary?
3. Press **2**: Change or Display an Abbreviations File? (You will now need to indicate that you plan to use the DBPA file.)
4. Press **1**: Change or Add Abbreviations?
5. Type the abbreviations (or symbols) and descriptions.
   a.   EN   ENGINEERS
   b.   AR   ARCHITECTS
   c.   DE   DESIGNERS
   d.   SP   SKILLED PERSONNEL/LABOR
6. Press **5**: Save This Version and Return (to main menu).

Access the screen to enter the resource data through the main menu:

1. Press **1**: Produce an Up-to-Date Plan?
2. For Do You Want to Work With: Press **DBP version 2** (or version that you are using).
3. In the data input screen, press **F2** to enter the resources data screen.
4. Line 1: (Showing node numbers 1,2) Type **EN2 DE6** [ENTER].
5. Line 2: (For node numbers 1,3) Type **EN6 AR4 DE12** [ENTER].
6. Continue until all resources are entered.

```
COMMAND?                          Current Plan: DBP      version 1

LINE   FROM   TO      RESOURCES ALLOCATED PER UNIT TIME           ?=HELP

 1 :    1  -> 2       EN2 DE6 TC2857 TC2857
 2 :       -> 3       EN6 AR4 DE12 TC4615 $C4615
 3 :       -> 4       EN4 DE4 TC2857 $C2857
 4 :    2  -> 3       EN1 DE2 SP10 TC3143 $C3143
 5 :    3  -> 7       EN3 SP20 TC10000 $C10000
 6 :    4  -> 5       EN4 DE8 TC2857 $C2857
 7 :       -> 6       EN2 DE4 TC1429 $C1429
 8 :       -> 8       AR4 DE8 TC2857 $C2857
 9 :    5  -> 6       EN2 DE1 TC816 $C816
10:     6  -> 7       EN4 SP16 TC4762 $C4762
11:     7  -> 11      EN5 SP20 TC10417 $C10417
12:     8  -> 9       AR2 DE2 TC1429 $C1429
13:     9  -> 10      AR2 SP12 TC3429 $C3429




 1DESCRP 2RESRCE 3INSERT 4NEW PG 5HDINGS 6GO TOP 7ZOOM  8BARCHT 9ESCAPE 10HELP
```

7. Press [ENTER], then [ESC] when all lines are completed.
8. Type **2** at *Enter a New Time Now Day Number* [ENTER].

Various resources reports can now be printed. For the daily Engineer's report:

1. Press **5**: Histogram for a Resource [ENTER].
2. Press **1**: That All Activities Start As Early As Possible? [ENTER].
3. Type **2** for The Day Number That You Want This Report to Start From.
4. Type **EN** for the Resource to be Analyzed in this Report: [ENTER].
5. Press **8**: Start or Cancel the Run.
6. Press [ENTER] to start the run.

Snyder Engineering begins an analysis of the requirements with the histogram (or load chart) reports. Analysis starting at the early stages of a project has a decided advantage. If any changes need to be made to the plan and schedule, they probably will not interfere much with the work in progress. One of the main concerns is the ability to make available the personnel/labor required for the Christopher Design/Build Project. Snyder Engineering may have several other design jobs in progress that may have an impact on the number of designers that can be made available over the 15 designers assigned.

Figure 6.9 shows the designer histogram (or load chart) for the Christopher Design/Build Project and includes the supply limit line that graphically portrays clearly the excess demand. The chart displays the early start schedule for the critical period October 2, 1995 through January 30, 1996. It is adequate for early resources planning, enough to alert Snyder Engineering of a potential problem. The firm needs to analyze the inconsistent pattern, consider some options, then decide on a course of action.

Period	Designers— Daily Requirements
2/Oct/95–16/Oct/95	22
16/Oct/95–13/Nov/95	18
13/Nov/95–27/Nov/95	34
27/Nov/95–11/Dec/95	28
11/Dec/95–25/Dec/95	22
25/Dec/95–1/Jan/96	18
1/Jan/96–15/Jan/96	6
15/Jan/96–30/Jan/96	2

This tabulation shows the heaviest demand is during the November 13, 1995 through December 24, 1995 period. Snyder Engineering had designated 15 designers for this project, yet as many as 34 designers are needed during a two-week period. Other high-demand situations are almost as disturbing. The heavy demand for designers close to the start of the project, for example, compounds the problem in three ways: (1) it creates a heavy cash flow burden at the outset of

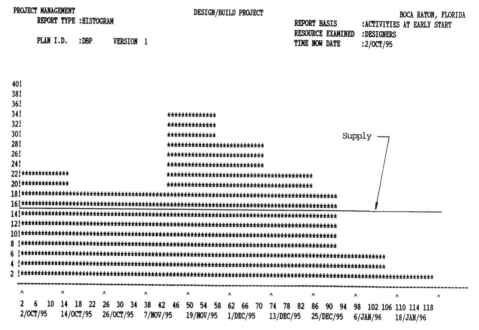

FIGURE 6.9   Computer-generated skilled personnel/labor load chart: Early start schedule from January 8, 1996 to the end of the project

the project, (2) it does not allow adequate time for the team to become acclimated to the project before being faced with heavy design assignments, and (3) it creates an immediate designer loading problem.

These problems need to be resolved at the beginning or at least the early stages of the project. Options to review include the following:

1. overtime for the designers during the heavy demand periods
2. temporary hiring of additional designers during the heavy demand periods
3. use of the late start schedule if it shows a more satisfactory demand for designers
4. examination of selected activities, preferably with high float values, where schedule changes would reduce the designer load during the heavy demand periods

For additional help in analyzing these options, Snyder Engineering can produce additional computer-generated charts. Figure 6.10 shows the designer loadings when the project activities are scheduled to begin at their latest start dates. This shows a five-week relief at the beginning of the project before the designer load increases to higher levels, although not as high as those of the earliest start schedule. When considering using the latest start schedule, the team must remember it has a tradeoff—in return for more reasonable resource loadings, all

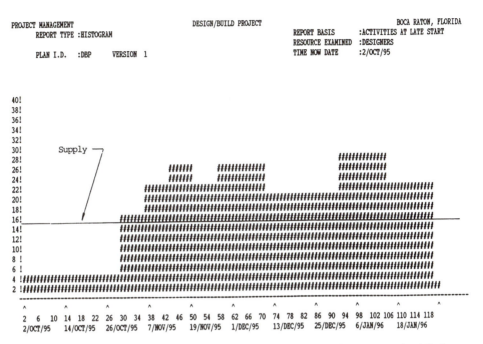

PROJECT MANAGEMENT   DESIGN/BUILD PROJECT   BOCA RATON, FLORIDA
        REPORT TYPE :HISTOGRAM   REPORT BASIS   :ACTIVITIES AT LATE START
                                                                       RESOURCE EXAMINED :DESIGNERS
        PLAN I.D.   :DBP   VERSION 1   TIME NOW DATE   :2/OCT/95

FIGURE 6.10   Computer-generated designer load chart: Late start schedule for
the Christopher Design/Build Project

the project activities are scheduled at their latest dates. A timing problem may
crop up as the project progresses as schedules for any activities "drift." The project end date changes proportionally with a change in any of its activities.

Concurrent with the designer load analysis, Snyder Engineering reviews the
skilled personnel/labor loads as they do show certain periods of heavy demand.
They are not as serious, since they exist at lower levels and at the latter stages of
the project. Figure 6.11 displays the skilled personnel/labor loads when using the
early start schedules. There is an additional demand, but for a relatively short
period, and it could be handled with additional overtime. While the latest start
schedule for the designers appears more desirable, it is not quite as suitable, as
shown in Figure 6.12. With this computer data available, Snyder Engineering can
begin analyzing the options.

The resource summary (a tabulated format) will also be helpful. It presents
the weekly summaries of the average daily labor/personnel requirements along
with costs for the same periods. Computer reports for both earliest start and latest start schedules are produced and compared. Some users may want to review
these summaries before generating the load charts. The report prepared on one
sheet can show the overall impact and pinpoint the trouble periods that can then
be highlighted with the load charts. These charts (Figure 6.13 for the early start
and Figure 6.14 for the late start schedule) display all the resources to be used on
the Christopher Design/Build Project.

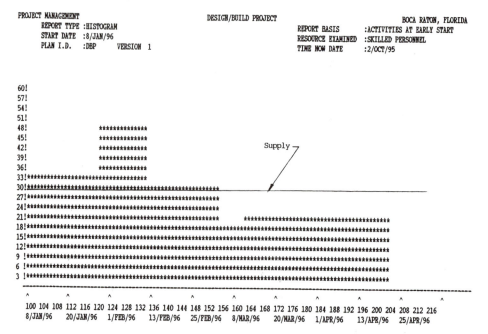

FIGURE 6.11 Computer-generated skilled personnel/labor load chart: Early start schedule from January 8, 1996 to end of project

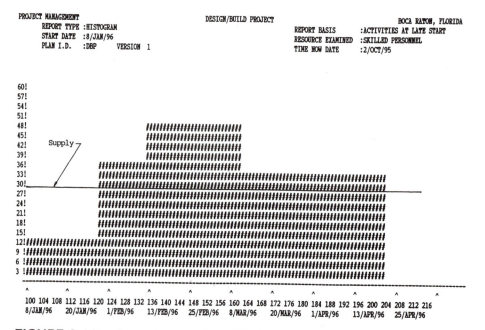

FIGURE 6.12 Computer-generated skilled personnel/labor load chart: Late start schedule from January 8, 1996 to end of project

| | EN=ENGINEERS | | DE=DESIGNERS | | TC=DAILY COSTS ($) | | $C=ACCUMULATED COSTS ($) |
| AR=ARCHITECTS | | SP=SKILLED PERSONNEL | | | | | |

	EN	DE	TC	$C	AR	SP
2/OCT/95	12!	22!	10329!	10329!	4!	!
9/OCT/95	12!	22!	10329!	10329!	4!	!
16/OCT/95	11!	18!	10615!	10615!	4!	10!
23/OCT/95	11!	18!	10615!	10615!	4!	10!
30/OCT/95	11!	18!	10615!	10615!	4!	10!
6/NOV/95	11!	18!	10615!	10615!	4!	10!
13/NOV/95	13!	34!	14901!	14901!	8!	10!
20/NOV/95	13!	34!	14901!	14901!	8!	10!
27/NOV/95	13!	28!	13473!	13473!	6!	10!
4/DEC/95	13!	28!	13473!	13473!	6!	10!
11/DEC/95	11!	21!	11432!	11432!	6!	10!
18/DEC/95	11!	21!	11432!	11432!	6!	10!
25/DEC/95	10!	17!	10289!	10289!	6!	12!
1/JAN/96	7!	5!	15674!	15674!	2!	32!
8/JAN/96	7!	5!	15674!	15674!	2!	32!
15/JAN/96	5!	1!	14245!	14245!	2!	32!
22/JAN/96	5!	1!	14245!	14245!	2!	32!
29/JAN/96	7!	!	18191!	18191!	2!	48!
5/FEB/96	7!	!	18191!	18191!	2!	48!
12/FEB/96	4!	!	81191!	8191!	2!	28!
19/FEB/96	4!	!	8191!	8191!	2!	28!
26/FEB/96	4!	!	8191!	8191!	2!	28!
4/MAR/96	4!	!	47621!	4762!	!	16!
11/MAR/96	5!	!	10417!	10417!	!	20!
18/MAR/96	5!	!	10417!	10417!	!	20!
25/MAR/96	5!	!	10417!	10417!	!	20!
1/APR/96	5!	!	10417!	10417!	!	20!
8/APR/96	5!	!	10417!	10417!	!	20!
15/APR/96	5!	!	10417!	10417!	!	20!
22/APR/96	!	!	!	!	!	!
29/APR/96	!	!	!	!	!	!
6/MAY/96	!	!	!	!	!	!
13/MAY/96	!	!	!	!	!	!
20/MAY/96	!	!	!	!	!	!

FIGURE 6.13 Computer-generated weekly summary of average daily resources using the early start schedule

```
PROJECT MANAGEMENT                              DESIGN/BUILD PROJECT                                    BOCA RATON, FLORIDA
         REPORT TYPE :COMPLETE RESOURCES REPORT                                    REPORT BASIS      :ACTIVITIES AT LATE START
                     :AVERAGE DAILY DEMAND PER WEEK                                SELECTION CRITERIA :ALL
         PLAN I.D.   :DBP        VERSION 1                                         TIME NOW DATE      :2/OCT/95
===================================================================================================================================
     EN=ENGINEERS                    DE=DESIGNERS              TC=DAILY COSTS ($)         $C=ACCUMULATED COSTS ($)
     AR=ARCHITECTS                   SP=SKILLED PERSONNEL
```

	EN	DE	TC	$C	AR	SP
2/OCT/95	4!	4!	2857!	2857!	!	!
9/OCT/95	4!	4!	2857!	2857!	!	!
16/OCT/95	4!	4!	2857!	2857!	!	!
23/OCT/95	4!	4!	2857!	2857!	!	!
30/OCT/95	10!	16!	7472!	7472!	4!	!
6/NOV/95	12!	22!	10329!	10329!	4!	!
13/NOV/95	12!	26!	10329!	10329!	4!	!
20/NOV/95	11!	22!	10615!	10615!	4!	10!
27/NOV/95	13!	26!	12044!	12044!	4!	10!
4/DEC/95	13!	26!	12044!	12044!	4!	10!
11/DEC/95	11!	19!	10003!	10003!	4!	10!
18/DEC/95	11!	19!	10003!	10003!	4!	10!
25/DEC/95	11!	19!	10003!	10003!	4!	10!
1/JAN/96	11!	27!	12860!	12860!	8!	10!
8/JAN/96	11!	27!	12860!	12860!	8!	10!
15/JAN/96	11!	21!	11432!	11432!	6!	10!
22/JAN/96	11!	21!	11432!	11432!	6!	10!
29/JAN/96	7!	2!	16191!	16191!	2!	36!
5/FEB/96	7!	2!	16191!	16191!	2!	36!
12/FEB/96	7!	!	18191!	18191!	2!	48!
19/FEB/96	7!	!	18191!	18191!	2!	48!
26/FEB/96	7!	!	18191!	18191!	2!	48!
4/MAR/96	7!	!	18191!	18191!	2!	48!
11/MAR/96	5!	!	13846!	13846!	2!	32!
18/MAR/96	5!	!	13846!	13846!	2!	32!
25/MAR/96	5!	!	13846!	13846!	2!	32!
1/APR/96	5!	!	13846!	13846!	2!	32!
8/APR/96	5!	!	13846!	13846!	2!	32!
15/APR/96	5!	!	13846!	13846!	2!	32!
22/APR/96	!	!	!	!	!	!
29/APR/96	!	!	!	!	!	!
6/MAY/96	!	!	!	!	!	!
13/MAY/96	!	!	!	!	!	!
20/MAY/96	!	!	!	!	!	!

FIGURE 6.14 Computer-generated weekly summary of average daily resources using the late start schedule

SUMMARY

A key aspect of the project planning process involves the resources required to complete the project. Planning resources is just as an important for many projects as scheduling and budgeting. Resources refer to the personnel, labor, equipment, material, and funding needed to complete a project. Disciplined resource planning involves mathematical calculations done either manually or with computer software that has special features to handle resource planning and leveling.

Resource planning is good business practice. It is used to reduce irregularities in employment. Consistent employment over the life of the project improves performance and maintains high morale among the employees and participants. One of the main objectives of resource planning is to level out personnel requirements by attempting to match supply with demand. The method used follows these steps:

- Specify the skills and/or resources required for each activity.
- Calculate (manually or by computer), by time period, demands of each skill and/or resource needed within that period. Use the earliest start schedule to determine necessary skills and/or resources.
- Compare the resources available for each activity with the number required for each period. If they compare favorably, then make no schedule adjustments.
- Leveling may be necessary where the required personnel and the available personnel differ. The intent of the leveling process is first to adjust the schedule of activities with the largest float values. It continues with activities having smaller float values, rearranging the schedule with each change.
- Leveling avoids adjustments to critical activities in the original schedule until there are no more float activities, and the project still shows personnel loading inconsistency.

The leveling process has a tradeoff: By reducing float values of affected activities, it increases the chances of the project not being completed as scheduled. The team needs to decide which is more important: maintaining the integrity of the original plan or schedule or matching resource availability to demand.

Manual leveling using bar chart time schedules for small projects will produce satisfactory answers. It may be time consuming, however, compared to using the resource allocation features of project management software. Comparing resource demands of the early start schedule with the late start schedule is one of the fundamental leveling methods to use. Advanced computer leveling programs consider activities individually until the total leveling process satisfies the supply and demand criteria.

Computerized scheduling combined with the project planning diagram is the basis for effective resource planning. Computerized resource planning allows for more effective leveling of the peaks and valleys of the labor and personnel demands than manual methods.

Exercises

1. Provide two reasons for using resource planning methods for the labor and personnel requirements of a project.
2. Using the project planning diagram in Figure 1.2, and the bar chart time schedule prepared for Chapter 3, Exercise 6:
   a. Prepare the weekly distribution for the average daily requirements of designers for the first 10 weeks of the project.
   b. Prepare a weekly distribution of the average skilled personnel requirements for weeks 10 through 20 of the project. Use the resources listed in the following table:

	Resources			
Activity	Architects (AR)	Engineers (EN)	Designers (DE)	Skilled Personnel (SP)
Structural steel engineering		1	4	
Fabricate and erect structural steel		2	4	
Construction engineering		6	12	
Phase I construction		6		15
Phase II construction		8		20
Equipment engineering		2	6	
Award contract		1		
Fabricate and deliver equipment		4		
Phase I installation		4		10
Phase II installation		4		10
Long-lead items engineering		4		
Long-lead items procurement		2		

3. Using computer software, the project shown in Figure 1.2, the resource data in Exercise 2, and the timing data developed in Chapter 3, Exercise 6:
   a. With the early start schedule, prepare the histogram (load chart) for designers (DE) and skilled personnel (SP) for the total project.
   b. From the prepared histograms, show the average daily requirements for designers and skilled personnel for the following periods of the project:

Period	Designers	Skilled Personnel
1 Nov. 95 – 31 Dec. 95		
1 Jan. 96 – 31 Mar. 96		
1 Apr. 96 – 30 Jun. 96		
1 Jul. 96 – End of Project		

4. Using computer software and the late start schedule, develop histograms for designers and skilled personnel.
5. Which schedule (early start or late start) seems more desirable to use for designers? Skilled personnel? Substantiate your findings by using the same periods from the table in Exercise 3 to complete the following tabulation:

	Average Daily Requirements			
Period	Designers		Skilled Personnel	
	Early Start	Late Start	Early Start	Late Start

a. What are the advantages and disadvantages of your selection?

Computerized Project Management

THIS CHAPTER WILL COVER

- Background of project management computer software

- Using computer software to
  a. Prepare screen reports
  b. Combine (merge) projects
  c. Produce reports based on the work breakdown structure (WBS)
  d. Produce the schedule with the precedence diagramming method

Using computer software productively for scheduling and controlling projects is the next step once you understand the basics of planning, scheduling, and controlling time, costs, personnel/labor, and other resources. The computer is the vital tool for successful project management. Persons committed to preparing projects depend on the computer for developing and analyzing time and cost schedules and for determining labor and personnel requirements. Its popularity for project use has resulted in a phenomenal growth of project management software packages. Today more than 200 project management computer programs are being marketed.

THE FUNCTION OF THE COMPUTER

The advantages of using computers are extensive. Computers make it possible to

- handle very rapidly the time and cost schedule calculations for a project that normally would require many more labor-hours if calculations were done manually
- handle accurately countless calculations that, if done manually, would be subject to errors and would require time to find and correct
- make project updating for control purposes effective as well as reduce the time-consuming and excessive costs of manual report preparation
- print project status reports in many varied readable formats

Computer programs use the data developed from the project planning efforts, principally from the project planning diagram. Most software will accept input data from either the arrow diagramming method or the precedence diagramming approach. Input data screens are structured for the activity-oriented items used in arrow diagramming or precedence diagramming showing the start-to-start, start-to-finish, finish-to-start, and finish-to-finish relationships with time delays and lags (see Figures 2.10 and 2.11). Both methods are illustrated in this chapter. While this book favors the arrow diagramming method, both methods lead to the same results and produce identical reports.

Most programs used in developing reports include several key features. First, they permit more than one activity to start at the beginning of the project and more than one activity to end at the end of the project. This is crucial when several projects are being controlled at the same time. Second, they include error-detection features. For example, the software with this book will not allow instructions for printing reports to proceed if the input is incorrectly entered or is incomplete.

Generally, the main properties of a project management software package include:

- **Project plan generation**: processes input data, builds the needed files, performs all necessary calculations, writes the project plan onto a master file, and prints any selected output reports.

- **Update**: updates an existing project plan by changing the input logic and timing data, prints the reported progress as requested, and records the updated data onto a master file.
- **Maintenance**: recomputes dates of project activities based on current or revised data, but does not alter the project logic plan.
- **Detailed reports**: varied reports are generated customized for the user. Reports can be printed in sequence based on the work breakdown structure for the specific user.

COMPUTER SOFTWARE PACKAGES

Deciding what computer software to use is a major concern. Normally, people involved in preparing project schedules will know little about the instructions to run the computer (the software) and the physical elements (hardware) that make up the computer. But users usually know what information they expect to receive from reports that the computer will generate.

A successful computer software supplier is one whose product package will suit the needs of the user and is user-friendly. It will be simple to use and learn. It should be designed to become productive as quickly as possible. A user-friendly software package should also display help functions on the screen, include understandable documentation, and be compatible with other business software.

The first-time user, who is no doubt a doer, cannot afford to spend too much time learning how the program works. A software supplier may need to give personal and direct instruction on the use of the product. It is not unusual for the supplier to demonstrate the product's performance by walking the user through at least one complete project applicable to the user's line of work.

A project evolves initially (with some computer help) when the user creates a project plan consisting of (1) objectives, (2) a work breakdown structure, and (3) the project planning diagram. Next, the user (1) develops time and cost estimates and (2) identifies the personnel/labor and other resources. (Software other than project management software may assist here, but experienced personnel normally supply estimates.) Developing all this information makes you think through your project.

When the computer becomes important is in preparing schedule reports. Using the computer for proper analysis of the reports generated encourages confidence that the project will be successfully completed.

Most software packages will allow at least 30 types of output reports to be printed or seen on the screen. These reports supply basic information on time, cost, and resource schedules; provide updates on the project status; and allow analysis of the project. Reports can follow various formats suited for all persons involved in the project. Reports most commonly used include

- **schedule reports**, which list the project activities with scheduled calendar dates, float values and durations

- **milestone reports**, which list milestones in the project plan and their scheduled dates
- **bar charts**, which show graphically the duration and relative time frame of all specified project activities
- **cost reports**, which show either graphically or in tabular form the daily costs and accumulated costs over a designated period
- **resources reports**, which show either graphically or in tabular form the labor/personnel and other designated resources over a designated time

USING PROJECT MANAGEMENT SOFTWARE

Demonstrations of the software used for the projects in this book are intended to illustrate its use in a relatively short time by focusing on hands-on applications. All fundamental packages, including this one, should come with a book that explains step-by-step application of planning, scheduling, and controlling projects.

Initial project management training emphasizes project planning. The use of the computer is subordinate during this time. The planning phase depends primarily on the deliberations and experience of the persons preparing the project detail. It becomes invaluable for all of the other phases. The computer completes countless calculations for preparing time, cost, and resources schedules as well as for controlling the project and evaluating project improvements.

Selecting project management software involves certain tradeoffs. Many available packages possess numerous bells and whistles—in this case, attractive screens and reporting printouts. Often, however, the extensive training to use such programs overshadows the main intent—to learn the basics of project management with the computer as a complementary tool.

Software should fulfill the above requirements for basic and fundamental use. It should also facilitate data inputting and report generation. It should be directly applicable to the teaching practices of this book or compatible with the specific instructional manual that is to be used. Instruction benefits from computer screens reproduced throughout the book replicating the menus and reports to be used.

As a recommended procedure for learning the use of computer software for the first time, this book employs essentially a step-by-step instruction approach. It also uses the same sample project, Christopher Design/Build Project, throughout the chapters (Chapters 2 through 7). This computer program is specifically designed for textbook study. It can analyze up to 75 activities, which is adequate for the purpose of learning. Larger projects would require an enlarged version of this program or a program more suitable to the specific project.

COMPUTER APPLICATIONS

This section covers instructions for performing several project management applications commonly used by those engaged in project management:

Using the computer screen to observe timing schedules and resource demands. This avoids printing out lengthy reports, thus saving time when analyzing information. It allows quick response time when analyzing what-if situations.

Merging two or more projects. More than one project may be going on at the same time. It would be worthwhile to know for planning purposes: (1) critical activities of all of the concurrent projects and (2) how, when distributed over a number of projects, the available resources compare to the total demand at any specific time.

Organizing a project using the WBS approach. The method is demonstrated with the sample project. The advanced case histories in this book use the WBS approach, and becoming familiar with it now should be worthwhile.

Use of precedence diagramming. This is a popular alternative method for constructing a planning diagram. The software with this book accepts this approach as well as the arrow diagramming method. Using the precedence diagram of the sample project in Chapter 2, this chapter runs through the input preparation for the reports.

Appendix A contains additional housekeeping features to enhance the user's understanding of this software.

Screen Reports

On-screen versions of standard reports, project bar charts, and project histograms can be generated from two locations: the data input screen and the report menu.

From the data input screen:

1. Press **F6** to go immediately into the screen bar chart. (Proceed with the same instructions shown in the reports menu for obtaining standard reports and histograms.)

When in the reports menu:

2. Press option **6:** Screen reports [ENTER]. Follow through with the same report ordering routines as the standard listings.

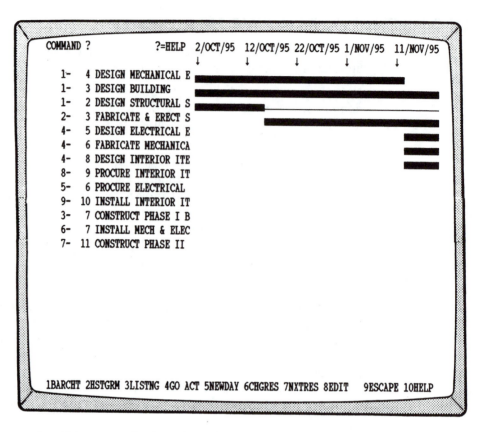

You should interpret the bar chart shown on your screen exactly as you would the bar chart that is printed out. The difference in appearance reflects the choice of characters used to denote critical activities, noncritical activities, and float. The screen report indicates critical activities by the highlighted bar, noncritical activities by the duller bar, and float by the hyphen (—). Activities with negative float (i.e., those with delayed scheduled finishes) are shown as flashing highlighted bars.

3. When the screen bar chart appears, press **F10** (Help) to get an understanding of using the screen reports. (For reference keep a copy of the
   help screen as you are using the screen reports.)

```
The COMMANDs Available are as Follows :-

Rght arrow- Scrolls the chart to the right.
Left arrow- Scrolls the chart to the left.
Down arrow- Moves the window down 18 lines.
Up   arrow- Moves the window up   18 lines.

B           - Type this command-to switch back to a barchart display.
R           - Type this to select a histogram for the period shown,
              you should enter a resource code if no resource is selected.
L           - Type this to select a normal date list for the activities.
C           - Use this command to change to a different resource code.
N           - This command will allow you to select a new start day number
              Enter the number or +/- x to go x days in either direction.
G           - This will allow you to goto another section of the plan by
              typing the first part of the description of the activity
              that you wish to have listed at the top of the page.
              Start with a single quote for a key search
ESC         - Type escape to leave the program.


Press any key to continue
                                                    9ESCAPE 10HELP
```

4. Press **F3** (Listing) to get the screen version of the standard listing report.

```
COMMAND ?            ?=HELP  EARLY START  LATE START   LATE FINISH   DUR  FLOAT

  1-   4 DESIGN MECHANICAL E   2/OCT/95     2/OCT/95    12/NOV/95    42   0
  1-   3 DESIGN BUILDING       2/OCT/95    30/OCT/95    28/JAN/96    91   28
  1-   2 DESIGN STRUCTURAL S - 2/OCT/95     6/NOV/95    19/NOV/95    14   35
  2-   3 FABRICATE & ERECT S  16/OCT/95    20/NOV/95    28/JAN/96    70   35
  4-   5 DESIGN ELECTRICAL E  13/NOV/95    13/NOV/95    10/DEC/95    28   0
  4-   6 FABRICATE MECHANICA  13/NOV/95    27/NOV/95    28/JAN/96    63   14
  4-   8 DESIGN INTERIOR ITE  13/NOV/95     1/JAN/96    14/JAN/96    14   49
  8-   9 PROCURE INTERIOR IT  27/NOV/95    15/JAN/96    11/FEB/96    28   49
  5-   6 PROCURE ELECTRICAL   11/DEC/95    11/DEC/95    28/JAN/96    49   0
  9-  10 INSTALL INTERIOR IT  25/DEC/95    12/FEB/96    21/APR/96    70   49
  3-   7 CONSTRUCT PHASE I B   1/JAN/96    29/JAN/96    10/MAR/96    42   28
  6-   7 INSTALL MECH & ELEC  29/JAN/96    29/JAN/96    10/MAR/96    42   0
  7-  11 CONSTRUCT PHASE II   11/MAR/96    11/MAR/96    21/APR/96    42   0




1BARCHT 2HSTGRM 3LISTNG 4GO ACT 5NEWDAY 6CHGRES 7NXTRES 8EDIT   9ESCAPE 10HELP
```

The listing becomes useful when you wish to view only specific dates and reduce the time or effort to generate complete printed reports. For example, if you wish to obtain just the late finish date of the last project activity to compare it with the project timing objectives, you would bring up the listing screen report.

5. Press **F2** (Histogram) to change the screen display to a resource histogram.
6. Enter the code **DE** for the Designer histogram.

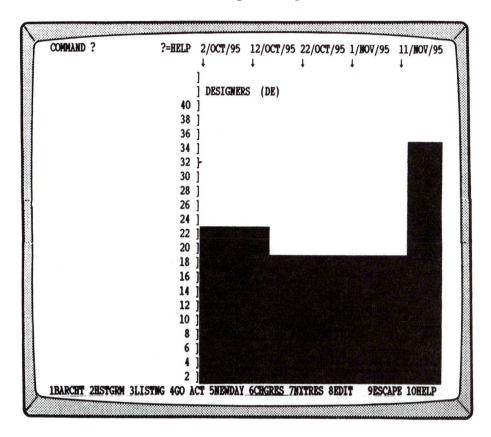

The screen will display about six weeks of the designer work load. For additional time, you can scroll the histogram using either **F5** (NEWDAY command) or the left and right cursor keys.

7. Press **F6** (Chrges) or **F7** (Nxtres).

8. Enter Resource Code **EN** for the Engineer histogram.

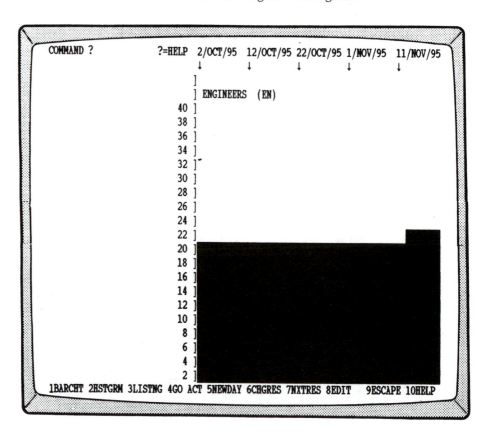

9. For the screen report of the daily project costs, press **F6** (Chrges) and
   enter Resource Code **TC**.

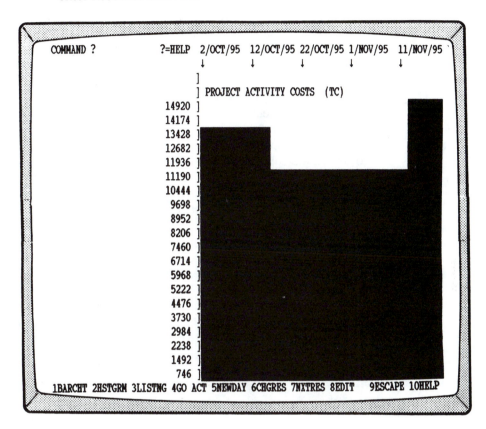

Press [ESC] to return to a dialogue to request additional reports or back to the
main menu.

Merging Plans

The computer's ability to merge several projects into one overall project is an important feature for several reasons. First, knowing all the critical items of the major projects that are going on at the same time is crucial to maintaining project control. Second, in the early stages of these projects it is desirable to know whether the available personnel will meet the requirements at peak periods. Several projects may have concurrent peak demands.

Let's assume that Snyder Engineering has taken on another project, Lynn Engineering Project, when the Christopher Design/Build Project is about one month underway. Snyder wants to know whether it will have an adequate supply of personnel, especially designers, to capably handle both projects. It is desirable to determine this as early as possible so Snyder can look at various options and make any adjustments to resolve the problem during the early stages of the project.

The first step is to combine, or merge, the two projects, simulating one large project, which the available software can do. The Christopher Design/Build Project and the Lynn Engineering Project will be combined and identified as the Engineer/Construct Project (code: DBPLYNN). The activities of the individual projects will take on their own identity in the reports with the addition of a prefix to each description: DBP for the Christopher Design/Build Project; LY for the Lynn Engineering Project. This type of categorizing permits sorting the project when using the alphabetical sequence in the report menu. This will allow each project team to review its own respective reports.

The merging process begins at the main menu screen:

1. Press option **4:** Do Some Housekeeping?
2. Press option **7:** Merge Plans Together?
3. Type **DBPLYNN** at the prompt *Please Enter the I.D. of the Plan to Be Created*.

4. Select **DBP** (and your choice of version number) at the prompt *Which Plan Do You Want to Merge Next.*
5. Type **50** [ENTER] at the prompt *Please Enter the Offset Value.*
6. Type **DB** [ENTER] at the prompt *Please Enter the Code to Be Inserted in Descriptions.*
7. Type **1** [ENTER] at the prompt *Please Enter the Position for the Code.*

```
┌─────────────────┐      ┌─────────┐      ┌─────────────────┐
│                 │      │IN CONTROL│      │                 │
│                 │      └─────────┘      │                 │
│                 │                       │                 │
└─────────────────┘                       └─────────────────┘

  Please enter the offset value : 50

  Press RETURN to add nothing to the node/activity numbers.


  Please enter the code to be inserted in descriptions:    DB

  Press RETURN if no code is to be inserted.


  Please enter the position for the code:  1

  Press RETURN to insert at the start.



                                              9ESCAPE 10HELP
```

8. Select **Lynn** at the prompt *Which Plan Do You Want Next*.
9. Repeating the steps: Type **10** [ENTER] at the prompt *Please Enter the Offset Value* [ENTER]; type **LY** at the prompt *Please Enter the Code to be Entered in the Description* [ENTER]; and type **1** at *Please Enter the Position for the Code* [ENTER].

```
                          ┌──────────────┐
 ┌──────────────────┐     │  IN CONTROL  │     ┌──────────────────┐
 │                  │     └──────────────┘     │                  │
 │                  │                          │                  │
 │                  │                          │                  │
 └──────────────────┘                          └──────────────────┘


 Please enter the offset value : 10

 Press RETURN to add nothing to the node/activity numbers.


 Please enter the code to be inserted in descriptions:    LY

 Press RETURN if no code is to be inserted.


 Please enter the position for the code:   1

 Press RETURN to insert at the start.




                                                 9ESCAPE 10HELP
```

10. Press **None of these** to indicate that no more plans are to be merged.

When the merging process is completed, the new plan will be saved and the main menu is displayed. Selecting the plan DBPLYNN and proceeding to the input screen show how the project is laid out. The following two screens show the results of an edit exercise that needs to be done before any reports can be generated that truly reflect the schedule.

```
COMMAND?                                  Current Plan: DBPLYNN   version 13

LINE   FROM    TO    DURATION   DESCRIPTION                          ?=HELP

 1 :  START-> 1                 NOT BEFORE DAY     2
 2 :        -> 21               NOT BEFORE DAY     32
 3 :   1   -> 2       14        DBP DESIGN STRUCTURAL STEEL
 4 :       -> 3       91        DBP DESIGN BUILDING
 5 :       -> 4       42        DBP DESIGN MECHANICAL EQUIPMENT
 6 :   2   -> 3       70        DBP FABRICATE & ERECT STRUCTURAL STEEL
 7 :   3   -> 7       42        DBP CONSTRUCT PHASE I BUILDING
 8 :   4   -> 5       28        DBP DESIGN ELECTRICAL EQUIPMENT
 9 :       -> 6       63        DBP FABRICATE MECHANICAL EQUIPMENT
10:        -> 8       14        DBP DESIGN INTERIOR ITEMS
11:    5   -> 6       49        DBP PROCURE ELECTRICAL EQUIPMENT
12:    6   -> 7       42        DBP INSTALL MECH & ELECT EQUIP
13:    7   -> 11      42        DBP CONSTRUCT PHASE II BUILDING
14:    8   -> 9       28        DBP PROCURE INTERIOR ITEMS
15:    9   -> 10      70        DBP INSTALL INTERIOR ITEMS
16:   10   -> 11      0         DUMMY
17:   11   ->FINISH             NOT AFTER  DAY     204
18:   21   -> 22      21        LY STRUCTURAL STEEL ENGINEERING
19:        -> 23      35        LY CONSTRUCTION ENGINEERING

 1DESCRP 2RESRCE 3INSERT 4NEW PG 5HDINGS 6GO TOP 7ZOOM   8BARCHT 9ESCAPE 10HELP
```

A continuation of the project is on the second screen.

```
COMMAND?                              Current Plan: DBPLYNN  version 13

LINE   FROM   TO   DURATION   DESCRIPTION                        ?=HELP

 1 :    21 -> 25     56       LY EQUIPMENT ENGINEERING
 2 :       -> 30     42       LY LONG LEAD ITEMS ENGINEERING
 3 :    22 -> 23     70       LY FABRICATE & ERECT STEEL
 4 :    23 -> 24     84       LY PHASE I CONSTRUCTION
 5 :    24 -> 27      0       DUMMY
 6 :       -> 29    140       LY PHASE II CONSTRUCTION
 7 :    25 -> 26      4       LY AWARD CONTRACT
 8 :    26 -> 27     84       LY EQUIPMENT FABRICATE & DELIVER
 9 :    27 -> 28     70       LY PHASE I INSTALLATION
10:     28 -> 29     56       LY PHASE II INSTALLATION
11:     29 ->FINISH           NOT AFTER DAY      346
12:     30 -> 31    126       LY LONG LEAD ITEMS PROCUREMENT
13:     31 -> 28      0       DUMMY




  1DESCRP 2RESRCE 3INSERT 4NEW PG 5HDINGS 6GO TOP 7ZOOM  8BARCHT 9ESCAPE 10HELP
```

The editing includes revising the START and FINISH activities of each of the two projects to their original dates. To position the correct dates, the calendar DBPC will need to be extended to cover the dates of the merged projects:

- Start date of Christopher Design/Build Project: 2
- Start date of the Lynn Engineering Project: 32
- Finish date of Christopher Design/Build Project: 204
- Finish date of the Lynn Engineering Project: 346

The identification screen may also need to be edited. Before producing reports, press **F5** (HEADINGS) to make any corrections or additions.

Using the alphabetical sequence option will produce a report that separates the Christopher Design/Build Project activities from the Lynn Engineering Project. Snyder Engineering may also generate the critical activities report showing floats of both projects.

Reports Using the Work Breakdown Structure

Using the WBS method, which permits an organizational approach to planning a project, highlights the responsibilities of each group involved in the project. The reports are arranged to combine the tasks of each group together, which clearly defines their roles.

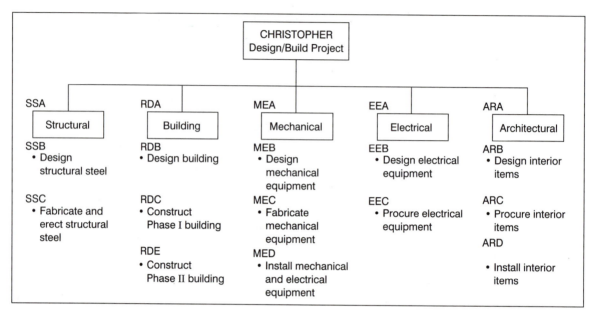

FIGURE 7.1   Coding the work breakdown structure for the Christopher Design/Build Project

Chapter 2 included an explanation of the WBS outline and how it was developed for the Christopher Design/Build Project. It not only is a great help in preparing the project plan, but it also becomes the source document for computer input data when using the alphabetical sequence in the report menu. An alphabetical letter is added to each task on the chart (as shown in Figure 7.1), then entered into the computer program. This coding allows organization of activities in the schedule and bar chart reports according to the responsibilities shown on the WBS chart.

Data are entered in the same manner as when preparing a new plan. Using the name DBPWBS for the new plan identifies it as the Christopher Design/Build Project. To identify it as the same Christopher Design/Build Project, the Identification screen can remain the same except for changing the left report title to Organization.

```
Please Enter Any Changes To the Following Information :-
----------------------------------------------------

Plan File Identification (1 to 8  Letters):  DBPWBS

Left  Hand Report Title  (Upto 30 Letters):  ORGANIZATION (WBS)

Centre    Report Title   (Upto 30 Letters):  DESIGN/BUILD PROJECT

Right Hand Report Title  (Upto 30 Letters):  BOCA RATON, FLORIDA

Calendar Code-Name       (1 to 8  Letters):  DBPC

Abbreviations Code-Name  (1 to 8  Letters):  DBPA

Plan Format Type    (A=Arrow, P=Precedence):  ARROW

Time Units per Day (Numeric,Upto 3 digits):  1


NOTE : Press RETURN only to Skip Item, or to Leave it Unchanged
       Press ESCAPE to Back Up One Step, or ? for HELP

                                          9ESCAPE 10HELP
```

Using the activities worksheet format described in Chapter 2 would simplify entering the input data from the WBS chart (Figure 7.1).

```
COMMAND?                              Current Plan: DBPWBS    version 13

LINE   FROM   TO   DURATION   DESCRIPTION                            ?=HELP

1 :    START-> 1                 NOT BEFORE DAY    2
2 :    1   -> 2       14         SSB Design structural steel
3 :        -> 3       91         BDB Design building
4 :        -> 4       42         MEB Design mechanical equipment
5 :        -> 7      147         MEA MECHANICAL   :WBS
6 :        -> 11     175         BDA BUILDING     :WBS
7 :        -> 31      84         SSA STRUCTURAL STEEL   :WBS
8 :    2   -> 3       70         SSC Fabricate & erect structural steel
9 :    3   -> 7       42         BDC Construct Phase I building
10:    4   -> 5       28         EEB Design electrical equipment
11:        -> 6       63         MEC Fabricate mechanical equipment
12:        -> 8       14         ARB Design interior items
13:        -> 10     112         ARA ARCHITECTURAL   :WBS
14:        -> 61      77         EEA ELECTRICAL   :WBS
15:    5   -> 6       49         EEC Procure electrical equipment
16:    6   -> 7       42         MED Install mech & elect equip
17:    7   -> 11      42         BDD Construct Phase II building
18:    8   -> 9       28         ARC Procure interior items
19:    9   -> 10      70         ARD Install interior items

 1DESCRP 2RESRCE 3INSERT 4NEW PG 5HDINGS 6GO TOP 7ZOOM   8BARCHT 9ESCAPE 10HELP
```

Follow the same procedure to get to the report menu:
1. Press **1:** Standard Report Listing.
2. Press **2:** Output in Alphabetical Sequence.

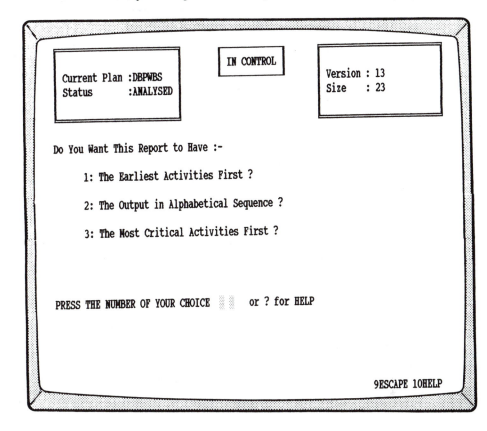

3. Type **11** at (1) *Starting Location* [ENTER].
4. Type **3** at (2) *Number of Characters* [ENTER].

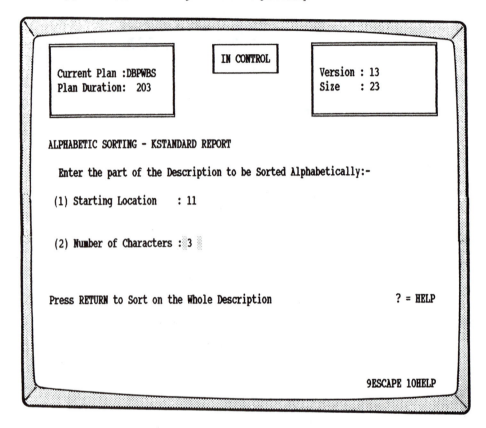

```
┌─────────────────────────────┐  ┌──────────────┐  ┌─────────────────────┐
│ Current Plan :DBPWBS         │  │  IN CONTROL  │  │ Version : 13        │
│ Plan Duration:  203          │  └──────────────┘  │ Size    : 23        │
└─────────────────────────────┘                     └─────────────────────┘

ALPHABETIC SORTING - KSTANDARD REPORT

  Enter the part of the Description to be Sorted Alphabetically:-

 (1) Starting Location    : 11


 (2) Number of Characters : 3


 Press RETURN to Sort on the Whole Description              ? = HELP



                                                   9ESCAPE 10HELP
```

5. Press [ENTER], [ENTER].
6. Press **8:** Start or Cancel the Run [ENTER]. Figure 7.2 shows the results.

```
PROJECT MANAGEMENT                              DESIGN/BUILD PROJECT                              BOCA RATON, FLORIDA
        REPORT TYPE :STANDARD LISTING                                   PRINTING SEQUENCE :Node/Activity Sequence
                                                                        SELECTION CRITERIA :ALL
        PLAN I.D.  :DBPWBS   VERSION  2                                 TIME NOW DATE     : 2/OCT/95
```

ACTIVITY DESCRIPTION	EARLIEST START	EARLIEST FINISH	LATEST START	LATEST FINISH	DURATION	FLOAT
4- 10 ARA ARCHITECTURAL :WBS	13/NOV/95	3/MAR/96	1/JAN/96	21/APR/96	112	49
4- 8 ARB Design interior items	13/NOV/95	26/NOV/95	1/JAN/96	14/JAN/96	14	49
8- 9 ARC Procure interior items	27/NOV/95	24/DEC/95	15/JAN/96	11/FEB/96	28	49
9- 10 ARD Install interior items	25/DEC/95	3/MAR/96	12/FEB/96	21/APR/96	70	49
1- 11 BDA BUILDING :WBS	2/OCT/95	24/MAR/96	30/OCT/95	21/APR/96	175	28
1- 3 BDB Design building	2/OCT/95	31/DEC/95	30/OCT/95	28/JAN/96	91	28
3- 7 BDC Construct Phase I building	1/JAN/96	11/FEB/96	29/JAN/96	10/MAR/96	42	28
7- 11 BDD Construct Phase II building	11/MAR/96	21/APR/96	11/MAR/96	21/APR/96	42	0 *
4- 61 EEA ELECTRICAL :WBS	13/NOV/95	28/JAN/96	13/NOV/95	28/JAN/96	77	0 *
4- 5 EEB Design electrical equipment	13/NOV/95	10/DEC/95	13/NOV/95	10/DEC/95	28	0 *
5- 6 EEC Procure electrical equipment	11/DEC/95	28/JAN/96	11/DEC/95	28/JAN/96	49	0 *
1- 7 MEA MECHANICAL :WBS	2/OCT/95	25/FEB/96	16/OCT/95	10/MAR/96	147	14
1- 4 MEB Design mechanical equipment	2/OCT/95	12/NOV/95	2/OCT/95	12/NOV/95	42	0 *
4- 6 MEC Fabricate mechanical equipment	13/NOV/95	14/JAN/96	27/NOV/95	28/JAN/96	63	14
6- 7 MED Install mech & elect equip	29/JAN/96	10/MAR/96	29/JAN/96	10/MAR/96	42	0 *
1- 31 SSA STRUCTURAL STEEL :WBS	2/OCT/95	24/DEC/95	6/NOV/95	28/JAN/96	84	35
1- 2 SSB Design structural steel	2/OCT/95	15/OCT/95	6/NOV/95	19/NOV/95	14	35
2- 3 SSC Fabricate & erect struct steel	16/OCT/95	24/DEC/95	20/NOV/95	28/JAN/96	70	35

FIGURE 7.2 Christopher Design/Build Project schedule based on work breakdown structure

Bar chart reports using WBS can also be produced by selecting the bar chart option in the report menu and following the same sequence as noted above. With the bar chart option, remember to use (0) when the date is requested. This is to ensure that the bar chart will be printed on one sheet. (Specifying the start date will produce a daily bar chart consisting of several sheets, which may be desirable for some purposes.)

Precedence Diagramming

Because both the arrow diagramming and precedence diagramming methods are the most popular diagramming styles used in the planning process, most project management software packages accept data for both. Chapter 2 has a section illustrating the use of precedence diagramming for the Christopher Design/Build Project. Figure 2.10 shows the precedence diagram for the Christopher Design/Build Project, which also serves as the source document for this inputting exercise.

1. Press **1:** Produce an Up-to-date Plan?
2. Type **DBPPREC** for the Plan File Identification.
3. Use the following screen as a guide when typing in the remaining information. (Note: Type **P** at *Plan Format Type for Precedence Diagramming.*)

```
Please Enter Any Changes To the Following Information :-
-------------------------------------------------------

   Plan File Identification (1 to 8  Letters):  DBPPREC

   Left  Hand Report Title  (Upto 30 Letters):  PRECEDENCE DIAGRAMMING

   Centre    Report Title  (Upto 30 Letters):  DESIGN/BUILD PROJECT

   Right Hand Report Title  (Upto 30 Letters):  BOCA RATON, FLORIDA

   Calendar Code-Name       (1 to 8  Letters):  DBPC

   Abbreviations Code-Name  (1 to 8  Letters):  DBPA

   Plan Format Type   (A=Arrow, P=Precedence):  PRECEDENCE

   Time Units per Day (Numeric,Upto 3 digits):  1


 NOTE : Press RETURN only to Skip Item, or to Leave it Unchanged
        Press ESCAPE to Back Up One Step, or ? for HELP

                                        9ESCAPE 10HELP
```

4. Type **Y** to get Input screen.
5. Press [ESC] to get required Start and Finish milestones.

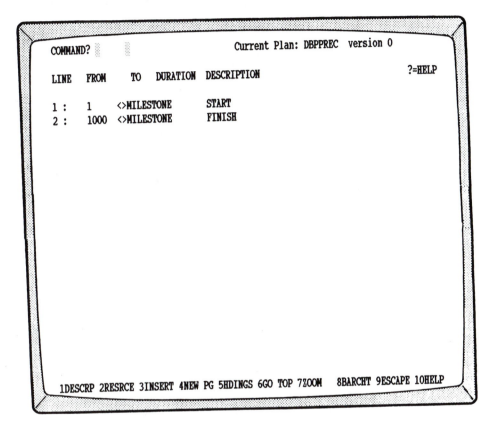

```
COMMAND?                          Current Plan: DBPPREC  version 0

LINE   FROM    TO   DURATION  DESCRIPTION                      ?=HELP

1 :    1     <>MILESTONE      START
2 :    1000  <>MILESTONE      FINISH












       1DESCRP 2RESRCE 3INSERT 4NEW PG 5HDINGS 6GO TOP 7ZOOM   8BARCHT 9ESCAPE 10HELP
```

6. Press **F3** (Insert) to start inputting data.
7. Press [ESC] when all data are entered .
8. Since additional data must be prepared and entered for precedence diagramming, a second page of data is required.

```
COMMAND?                          Current Plan: DBPPREC  version 2

LINE    FROM    TO    DURATION   DESCRIPTION                        ?=HELP

1 :     1     <>MILESTONE        START
2 :           --> 2              NOT BEFORE DAY     2
3 :     2             14         DESIGN STRUCTURAL STEEL
4 :           --> 3   0             FINISH->START
5 :           --> 9   0             START ->START
6 :           --> 10  0             START ->START
7 :     3             70         FABRICATE & ERECT STRUCTURAL STEEL
8 :           --> 4   0             FINISH->START
9 :     4             42         CONSTRUCT PHASE I
10:           --> 8   0             FINISH->START
11:     5             28         DESIGN ELECTRICAL EQUIPMENT
12:           --> 6   0             FINISH->START
13:     6             49         PROCURE ELECTRICAL EQUIPMENT
14:           --> 7   0             FINISH->START
15:     7             42         INSTALL MECHANICAL & ELECTRICAL EQUIP
16:           --> 8   0             FINISH->START
17:     8             42         CONSTRUCT PHASE II BUILDING
18:           --> FINISH          UNSCHEDULED
19:     9             91         DESIGN BUILDING

1DESCRP 2RESRCE 3INSERT 4NEW PG 5HDINGS 6GO TOP 7ZOOM  8BARCHT 9ESCAPE 10HELP
```

```
 COMMAND?                              Current Plan: DBPPREC  version 2

 LINE   FROM    TO   DURATION  DESCRIPTION                          ?=HELP

  1 :    9               91    DESIGN BUILDING
  2 :         --> 4       0        FINISH->START
  3 :    10              42    DESIGN MECHANICAL EQUIPMENT
  4 :         --> 5       0        FINISH->START
  5 :         --> 11      0        FINISH->START
  6 :         --> 12      0        FINISH->START
  7 :    11              63    FABRICATE MECHANICAL EQUIPMENT
  8 :         --> 7       0        FINISH->START
  9 :    12              14    DESIGN INTERIOR ITEMS
 10:         --> 13      0        FINISH->START
 11:    13              28    PROCURE INTERIOR ITEMS
 12:         --> 14      0        FINISH->START
 13:    14              70    INSTALL INTERIOR ITEMS
 14:         --> 8       0        FINISH->FINISH
 15:    1000 <>MILESTONE      FINISH



    1DESCRP 2RESRCE 3INSERT 4NEW PG 5HDINGS 6GO TOP 7ZOOM   8BARCHT 9ESCAPE 10HELP
```

9. Type **2** [ENTER].

Return to the main menu and select the report option to get reports. These reports will have the same format as those produced using the arrow diagramming method.

SUMMARY

Using the computer for project management affords the following benefits:
1. Computers very rapidly analyze the scheduling and costs associated with a project that would normally require many times more labor-hours if calculations were done manually.
2. They accurately handle countless calculations that would otherwise be subject to errors.
3. They print project status and results in useful and readable formats.
4. Computers are invaluable for analysis and simulation studies.

Well over 200 software packages focus exclusively on project management, and most of them have the ability to transfer data to other types of software packages. Choosing the right software is a joint effort with the supplier.

Software cannot, on its own, create project objectives, design a work breakdown structure and the project planning diagram showing all activities and their relationships, develop time and cost estimates, and identify the personnel/labor and other resources for the project activities. These require thought and effort by the user, which in turn require thinking through the project. On the other hand, software contributes heavily to preparing schedules for timing, costs, and other resources. Software also enables project analysis, which, when combined with all of the other features, leads to project success.

The software used for the projects in this book can be understood and put to use in minimal time. It is a fundamental package with no sophisticated features requiring extensive training. Its features include a screen that can be used handily for analysis and simulation studies to improve the project performance.

Reports generated by the computer include (1) a listing of the starting and finishing scheduled dates of the project activities, (2) a listing of critical project activities, (3) float times of the project activities, (4) cost schedules, and (5) resource loadings.

Exercises

1. What are the advantages of using project management software to schedule a project?
2. Refer to Figure 1.2 planning diagram and the durations of these activities:

Activity	Weeks
Structural engineering	3
Fabricate and erect structural steel	10
Construction engineering	5
Phase I construction	12
Phase II construction	20
Award contract	4
Equipment engineering	8
Fabricate and deliver equipment	12
Phase I installation	10
Phase II installation	8
Long-lead items—engineering	6
Long-lead items—procurement	18

Prepare a computerized milestone report showing the following events:
- start project
- project complete
- start *phase I construction*
- start *phase II construction*
- complete *fabricate and deliver equipment*
- complete *phase I Installation*; start *phase II*

3. Construct a work breakdown structure for the project shown in Exercise 2.

   Structural
   Structural steel engineering
   Fabricate and erect steel

   Construction
   Construction engineering
   Phase I construction
   Phase II construction

   Equipment
   Equipment engineering
   Award contract
   Fabricate and deliver equipment
   Phase I installation
   Phase II installation
   Long-lead engineering
   Long-lead procurement

4. Using the WBS for the project in Exercise 2, prepare a computerized schedule showing earliest start and finish, latest start and finish, duration, and float of each project item.

5. Generate a computerized bar chart (early start schedule) for the project in Exercise 4.

CASE HISTORIES

C H A P T E R   8

An Architectural/ Engineering Project: Planning CADD and Engineering Personnel Requirements

GAI is an architectural/engineering (A&E) firm formed in 1982 with eight employees offering specialized services in computer-aided drafting and computer-aided design (sometimes designated as CADD), a new field when the firm began. Since then, GAI has grown into one of the largest A&E firms in the midwestern United States.

Over the years, GAI has expanded beyond its initial CADD services to include traditional architecture and engineering (A&E) services. More recently, the services have expanded further to reflect clients' needs:

- engineering—civil, structural, mechanical, electrical, environmental, industrial/manufacturing
- architecture—interior design, landscape architecture
- plant facilities—plant/process layout, computerized facilities management, operations and maintenance
- computer services—CADD, computer-aided engineering, CADD input and scanning, computer programming, automated mapping
- expanded services—project management, design/build, construction management

The normal complement of GAI consists of 150 engineers, designers, and architects; 96 drafters, CADD operators, and computer programmers; and 27 administrative and clerical personnel. GAI operates one of the largest computer systems for A/E purposes in the Midwest. Since the number of terminals

173

and work stations in a system is less important than how the system is used, GAI considers its computer capability rather high for the range of services offered. These computer systems provide compatibility, versatility, standardization, and control over a broad range of computer applications, including CADD services, specifications, engineering calculations, desktop publishing, database development, and scanning.

CADD SERVICES

GAI offers CADD services through Microstation and AutoCAD—two of the most popular and powerful CADD systems available today. As one of the earliest users of CADD in the A&E profession, GAI has now expanded its CADD capabilities to provide

- enhanced architecture and engineering capabilities by computerizing specifications, engineering calculations, and construction drawings
- computer-assisted facility management databases that allow managers to maintain up-to-date facility plans and asset reports
- consulting services, including the establishment of procedures for systems operations/maintenance, and user training
- presentation materials, such as three-dimensional graphics and computerized renderings
- document conversion services, such as the scanning or conversion of manually produced A/E drawings
- software customization provided by an in-house programming staff

GAI can also provide drawings and data compatible with virtually all major CADD systems. GAI's computer network allows its systems to share dissimilar data and peripheral equipment. The network incorporates a variety of systems, including Digital VAX computers, MS-DOS and Windows (including Windows 95) personal computers, and UNIX work stations and servers.

TOTAL QUALITY MANAGEMENT

GAI has a total quality management (TQM) program with senior managers and employees from production and administrative groups forming a participative management committee. The TQM committee looks for ways to improve operations by building quality into the company's services.

- Regular in-house progress meetings check that the client's project criteria are met and that coordination between disciplines is effective.
- Performance assessment surveys are distributed to clients at project end. Employees are recognized for positive comments. Any problems are studied, and procedural changes are made to improve performance.

- An emphasis on automating design and document preparation takes advantage of GAI's strong computer capabilities.
- Employees are involved through company-wide meetings and a monthly newsletter. Also, skill development is encouraged through in-house training, seminars, and a tuition reimbursement program.

PROJECT BACKGROUND

On January 16, 1996, FGD, a large parts supplier, aware of GAI's performance and reputation, invited the firm to submit a proposal to design a fabrication facility having a roof area of approximately 350,000 square feet with the potential to expand to 500,000 square feet. Only large business and industrial firms can now support their own design groups, and since FGD is not that large, it engaged the services of reputable A&E firms to design any new buildings, building additions, or building modifications it needed.

The facility will be located on a site that FGD has yet to acquire—it expects the selected A&E firm to assist in the site selection. FGD is forecasting the facility to be completed no later than 24 months from the date of first contact with GAI (January 16, 1996). Other important milestones that GAI needs to consider in its proposal:

- Purchase site by May 1, 1996.
- Complete site engineering and design by June 15, 1996.
- Schedule groundbreaking ceremonies for August 16, 1996—the official start of site construction.
- Complete building design by November 15, 1996.
- Building construction should start no later than December, 1996.

For financial security reasons, FGD would not reveal the total cost of the facility, but did indicate to GAI that $18 million has been allocated for the site and building costs, including engineering fees. These fees represent GAI's pricing proposal for the engineering and design hours, material and equipment expenses needed to produce the drawings, indirect (overhead) costs, and profit.

FGD briefed GAI on the initial concepts of the project. To assist in the proposal preparation they gave GAI a sketch of the building layout and a *facility description*—a brief summary of the new facility (see Appendix B). Additional information provided GAI to help in preparing a proposal included the objectives, expectations, and limitations for the proposed facility.

Scope of Services. The principal services expected shall include:

1. Assist in preparing design concepts—guidelines and procedures.
2. Prepare preliminary designs and layouts.
3. Prepare final design drawings and specifications.

Additional services include:

1. Prepare and issue proposal forms (with drawings and specifications).
2. Write and issue bulletins for added and deleted work.
3. Estimate cost of construction.
4. Provide rendering assistance to contractors and equipment suppliers during proposal preparation period.
5. Review all shop and detail drawings to determine their compliance with the contract drawings and specifications.
6. Maintain tracings of the construction drawings to reflect the as-built condition of the work.
7. Provide assistance to the contractor and equipment supplier during course of construction by furnishing sketches.

Codes and Regulations.   GAI will prepare the designs, plans, and specifications in accordance with all federal, state, county, and local building and environmental regulations.

Owner's Representative.   All building construction and/or equipment installation work under contract will be handled for and on behalf of FGD by a representative of GAI. If FGD elects to use a construction manager, GAI will prepare a separate set of instructions to specify the construction manager's duties and responsibilities. (The construction manager provision will be shown in the engineering proposal as an alternate item.)

PREPARING THE ENGINEERING PROPOSAL

To prepare a detailed proposal to support the fee (which can range from 6 percent to 15 percent for designing this type of a facility project), GAI will need to develop more background information. The largest cost item will be the total direct engineering and design hours—to help substantiate these costs, GAI will prepare a plan and schedule for the project. This is a preliminary plan and schedule, as most of the project details are not known until the facility design is further developed. (Once awarded the design contract, GAI expects to expand it into a master plan and schedule to manage engineering and design work.)

GAI will develop a proposal according to this step-by-step approach:

1. Prepare a work breakdown structure (WBS) that outlines the work or activities of each design phase—concepts, preliminary, and final.
2. Construct a summary planning diagram.
3. Allot the personnel necessary to complete each of the activities.
4. Generate a computerized schedule to satisfy the FGD milestones.
5. Produce a computerized personnel loading tabulation that displays the total engineering and CADD hours to be expended on the FGD Project.

The total direct personnel costs are the product of multiplying the engineering and CADD hours by GAI's composite labor rates. These rates include the acceptable overhead and profit. Added to these costs will be charges for materials, travel, meeting expenses, and other overhead. Pricing for these items will be based on prior experience on similar jobs.

For a better understanding of the total project, GAI will divide it into three segments—design concepts, preliminary design, and final design—in constructing a cost proposal.

Design Concepts Phase

The initial design phase delineates the basic site, building, and administration criteria for the project. FGD furnishes most of the required information with GAI engineering and CADD assistance. To understand its requirements, GAI does prepare a WBS to help in developing the engineering proposal. As shown in Figure 8.1, the WBS is an excellent way to display all of the work involved in completing this phase. The WBS divides the work into manageable activities—the first step in preparing the planning diagram.

The main objective of the design concepts phase is to define the basic requirements of the site, building, and administration of the project. FGD will be responsible for acquiring the site, but will require some engineering and CADD assistance from GAI. The planning diagram will show the division of responsibility.

To assist in site selection, FGD furnishes site requirements prepared in the form of a report. Its contents include a site layout, building size, and configuration (prepared by GAI CADD personnel under FGD direction), consumption and type of utilities, environmental data, facility population data, and transportation data comprising incoming and outgoing freight, traffic control, and flow. FGD furnishes this report to a firm specializing in site investigation (or to qualified persons in their own firm). The firm or persons will then use it as a guide to identify sites fitting the requirements. The site investigation process uses a checklist when reviewing individual sites and for making comparisons of the different sites. (See Appendix B for a typical report.) Comparisons may be summarized in a matrix showing the major tangible and intangible factors that are "weighted" according to their importance. The tangible factors are the costs to construct and operate the facility. Intangible factors include labor supply and climate, attitudes toward business, and quality of life—schools, cultural activities, residential areas, and so forth.

FGD receives for review a report identifying the selection of two or three favored sites. After FGD management reviews and visits these favored sites (signing purchase options on them to avoid price escalation or to ensure ownership), one will be selected. The purchase negotiation process can be a lengthy one. It consists of meetings with the property owner and with government officials concerning price, tax abatements, training benefits, and other issues.

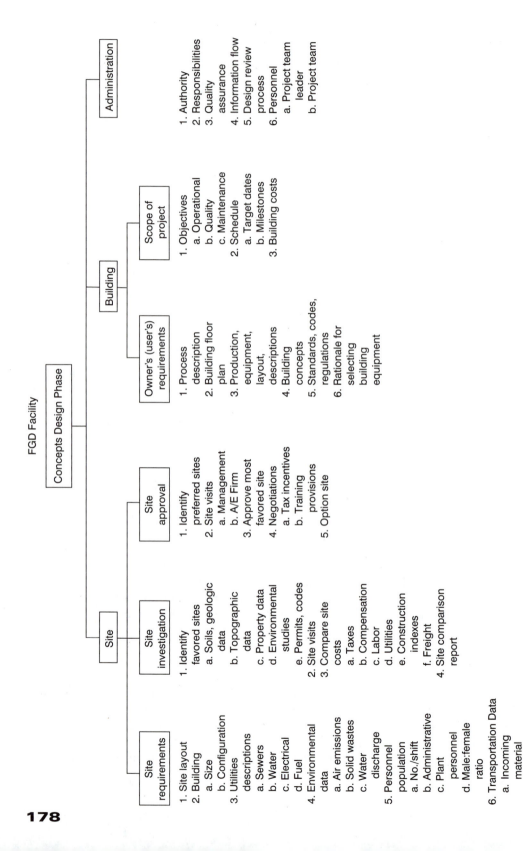

FIGURE 8.1 Design concepts phase of the FGD facility project

Concurrent with the site investigation, GAI reviews the owner's or user's requirements (prepared, in most part, by FGD). These include a description of the process, building floor plan and equipment layout (prepared by GAI CADD personnel under FGD direction), equipment loadings and dimensions, and special building features.

The user's requirements is a most important document, providing the basis for the engineering proposal. Several meetings will be needed to clarify and possibly modify the document after GAI personnel review it.

Once GAI has completed the plan for the design concept phase, the firm prepares a planning diagram showing all activities and their durations (Figure 8.2). Using this diagram and the manual scheduling method, this phase can be completed in 13 weeks. With contingency allowances, FGD's requirement of four months to complete this phase is plausible.

Preliminary Design Phase

Preliminary design is essentially a continuation of design work associated with the site and building. Site design and a detailed construction study make up most of the work in this phase. The WBS in Figure 8.3 shows the extent of the work that must be done to complete the preliminary design phase. For this project, FGD elects to "fast track" the site, meaning it will complete the site design and start site construction while continuing to work on the building design. The objective is to start site construction four months from the time GAI was authorized to proceed with the site design, for which limited funds have been advanced. (FGD has also authorized GAI to begin a detailed construction study report, which is a prerequisite to working on the final building designs.)

The final site design drawings and specifications include the building configuration and location on the site, earthwork and grading requirements for a level site, roadways, access to main roads with traffic control layouts, and underground mechanical and electrical utilities. Prior to completing these designs, GAI will conduct soils and geologic tests to ascertain the subsurface conditions and topographic surveys needed to determine the extent of the earthwork and grading requirements.

The detailed construction study is intended to present the preliminary building designs in an organized fashion. The scope of work is divided into the characteristic sections of an engineering project: architectural, civil, structural, mechanical, and electrical. Included will be architectural and structural steel layouts, schematics for the mechanical systems (including HVAC, utilities), and the electrical systems. (Appendix C shows a typical detailed construction study report.)

FGD and GAI jointly will be developing the quality standards to be incorporated into the final design and specifications for materials, products, and quality

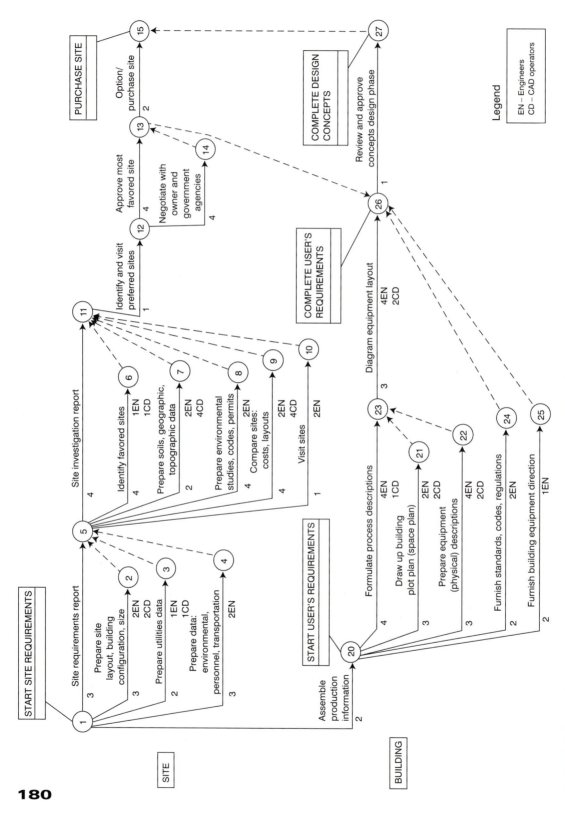

FIGURE 8.2 Planning diagram/design concepts phase for the FGD facility project

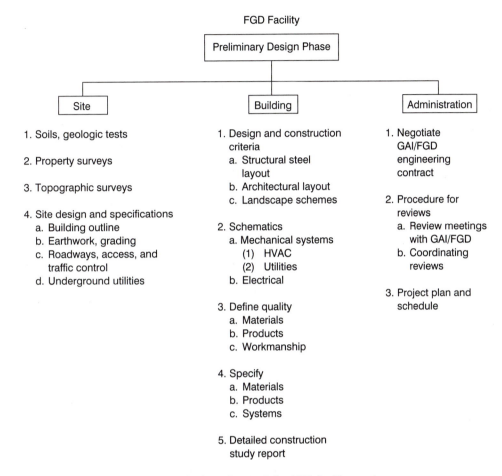

FGD Facility

Preliminary Design Phase

Site

1. Soils, geologic tests

2. Property surveys

3. Topographic surveys

4. Site design and specifications
   a. Building outline
   b. Earthwork, grading
   c. Roadways, access, and traffic control
   d. Underground utilities

Building

1. Design and construction criteria
   a. Structural steel layout
   b. Architectural layout
   c. Landscape schemes

2. Schematics
   a. Mechanical systems
      (1) HVAC
      (2) Utilities
   b. Electrical

3. Define quality
   a. Materials
   b. Products
   c. Workmanship

4. Specify
   a. Materials
   b. Products
   c. Systems

5. Detailed construction study report

Administration

1. Negotiate GAI/FGD engineering contract

2. Procedure for reviews
   a. Review meetings with GAI/FGD
   b. Coordinating reviews

3. Project plan and schedule

FIGURE 8.3 Preliminary design phase of the FGD facility project

standards. The two firms will also be supplying specific brand names or equivalents for materials to be included in the final design and specifications.

During this phase, the two firms will discuss how to handle design review meetings—their location, subject matter, and frequency. Although no formal engineering agreement exists, GAI needs to know these arrangements to estimate personnel and other matters on meeting attendance.

Concurrent with the preliminary design work, GAI will be developing the project plan and schedule, including the allocation of the engineering and CADD personnel needed to complete each activity. During this phase, GAI can complete its proposal. Assuming successful contract negotiations, the contract is approved and GAI receives notice to proceed to the final design phase.

Figure 8.4 shows the planning diagram for the preliminary design phase. This phase can be completed in 13 weeks, within the time allowed by FGD.

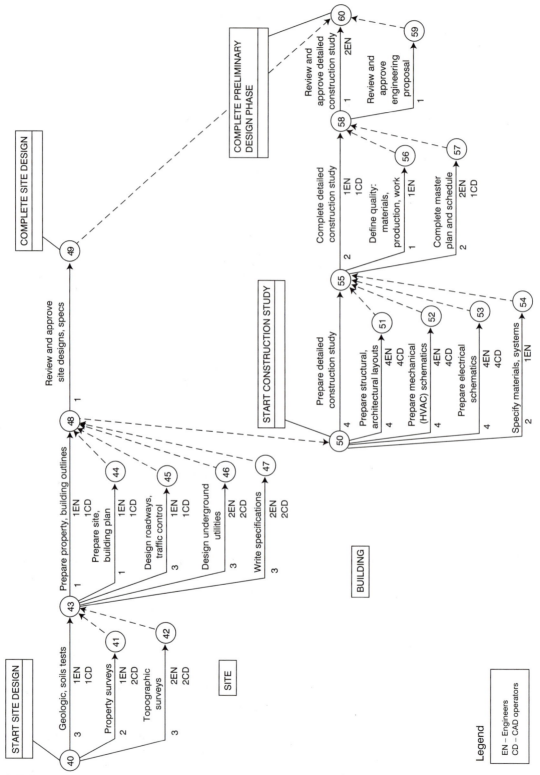

FIGURE 8.4 Planning diagram/preliminary design phase for the FGD facility project

Final Design Phase

Figure 8.5 shows the WBS for the final design phase in the FGD project. When GAI completes this phase the site construction contractor will essentially have the site ready for the start of building construction. By the end of this phase the process of selecting the general building contractor will be in its final stages.

Within this phase GAI completed the site bid documents, advertised for bidders, reviewed the bids, and (with FGD approval) awarded the contract to the contractor with the lowest and best proposal. The site work (also known as site improvements) continued during the final phase.

Concurrently GAI was preparing the contract documents consisting in part of a complete set of building designs and specifications. These are clearly defined by the characteristic sections of a design project: structural steel, architectural, landscape, mechanical (HVAC, utilities), and electrical.

In addition to the construction drawings and specifications, GAI prepares the remaining set of contract documents that consist of the following sections:

- proposal (bid) form (Figure 8.6)
- general conditions for building construction (Figure 8.7)
- special requirements (Figure 8.8)

The proposal form will refer to that which the contract authorizes and for which funds have been approved. A proposal form will usually include a unit price schedule (to be used for additions and deductions from the base bid proposal),

---

FGD Facility

Final Design Phase

Site

1. Bid documents
2. Contract award
3. Site construction

Building

1. Final design and drawing specifications
   a. Structural steel
   b. Architectural
   c. Landscape
   d. Mechanical systems
      (1) HVAC
      (2) Utilities
   e. Electrical
2. Bid forms
   a. Advertisement—invitation for bids
   b. Instructions to bidders
   c. Proposal form
   d. Agreement form—between owner and contractor
3. Bid documents
   a. Special requirements
   b. General conditions

Administration

1. Coordinate design reviews
   a. Review meetings
   b. Contacts
2. Final decision approvals

FIGURE 8.5 Final design phase for the FGD facility project

PF-1
FORM OF PROPOSAL

NAME OF BIDDER_____

ADDRESS_____PHONE _____

DATE _____

PF.01   PROPOSAL:

The Undersigned, hereinafter referred to as the Contractor, proposes to provide the complete design and furnish all labor, materials, tools, equipment, and supervision required and to perform all work required for materials and equipment procurement and installation in connection with the FGD Facility Project, for the FGD Company, hereinafter referred to as the Owner, in strict accordance with the Owner's General Conditions, Drawings, and Specifications as prepared by GAI, Inc. for the lump sum firm price of _____ Dollars ($ _____).

PF.01A   SCHEDULE OF ALTERNATES:

The Undersigned submits the following alternate prices, giving the amounts to be added or deducted from the base lump sum proposal amount. The alternate prices shall include all charges for incidental expenses, supervision, taxes, insurance, overhead, and profit.

Alternate A:

_____ $_____

_____

_____

Alternate B:

_____ $_____

_____

_____

Alternate C:

_____ $_____

_____

_____

PF.02   TIME FOR COMPLETION:

The Undersigned agrees to begin work immediately after notice of award of contract and to complete all work shown on the Drawings in _____ consecutive calendar days, Sundays and Holidays included, after date of notice of award of contract.

All engineering, construction, and fabrication work, including procurement of purchased components and subcontractor work, if applicable, shall be scheduled so as to complete construction on or before the completion date noted below:

Construction Completion: _____

FIGURE 8.6   Sample proposal form included in the contract documents

1. Definitions
2. Examination of premises
3. Surveys
4. Laws, ordinances, and regulations
5. Building permits
6. Taxes
7. Taxes: assigned orders or contracts
8. Alternate, separate, and unit prices
9. Acceptance and rejection of proposals
10. Specifications
11. Specifications and drawings to be cooperative
12. Signed plans and specifications
13. Number of working drawings and specifications
14. Assignment and subletting of contract
15. Review of contractor's drawings
16. Owner's options
17. Approval of equipment and material manufacturers
18. Samples to be submitted
19. Tests
20. Measurement and fitting of parts
21. Inspection of work away from premises
22. Quality of materials and labor
23. Delivery of materials
24. Removal of unfit units
25. Moving materials
26. Accident prevention
27. Explosives
28. Fire precautions and protection
29. Liability insurance
30. Owner's and contractor's responsibilities: fire and certain other risks
31. Contractor's responsibility for personal injuries and property damage
32. Contractor's responsibility: other risks
33. Progress schedule and time of completion
34. Contractor responsible for coordination and quality of work
35. Priority of items of work
36. Delays and extensions of time
37. Acceleration of work
38. Contractor's superintendent
39. Assistance by resident engineer or architect-engineer
40. Cooperation
41. Contractor to assist owner
42. Contractor's meetings
43. Installation of owner's equipment and machines
44. Alterations and additions
45. Patents
46. Performance bond
47. Liens
48. Schedule of prices and allocation of owner's cost
49. Contractor's payment requests
50. Payments to contractor
51. Patching and replacing of damaged work
52. Glass damage
53. Contractor's default
54. Suspension of operations
55. Termination by owner
56. Cleaning of premises
57. Guarantee

FIGURE 8.7   General conditions items included in the contract documents

100.00-1
SPECIAL REQUIREMENTS

100.1   <u>GENERAL NOTE:</u>

The provisions of Gilmore Industries "General Conditions for Lump Sum Contracts" bound herewith form a part of the following Specifications, and the Contractor shall consult them in detail for instructions pertaining to the work.

In the event of a conflict between provisions of the "General Conditions" and provisions of the "Special Requirements," those of the "Special Requirements" will take precedence.

100.02   <u>DRAWINGS AND DATA SHEETS:</u>

The accompanying drawings and the attached data sheets form a part of these Specifications and all work mentioned or indicated thereon in any manner shall be performed as though it were written out and described under the various headings of the Specifications.

100.03   <u>SCOPE OF WORK:</u>

The work contemplated by these Specifications consists of furnishing all labor, material, equipment, and services required for providing a complete facility for the FGD Company. The system shall provide monitoring, indication, and operation of equipment and systems as hereinafter specified and scheduled and shall include all accessories, appurtenances, and incidental items required for the completion of same, even though such items are not specifically shown or mentioned herein. It is the intent of these Specifications to provide a system that will accomplish not only those functions required at this time, but shall also permit future additions.

Contractor shall be responsible for, but not limited to, the following:

1. The furnishing of all materials, equipment, and labor of every kind required to carry out the intent of the Drawings and Specifications. Where it is not specifically indicated, materials and equipment are to be furnished by this Contractor.

2. The quality and sufficiency of all materials so furnished.

3. All mechanical and control piping, conduit, wire and cable, panels, etc., necessary to provide a completed installation.

4. Preparation of all drawings, details, and specifications necessary for the installation and connection of all equipment.

5. Completion of all systems as required to insure successful completion of the project.

FIGURE 8.8   Sample page of special requirements items included in the contract documents

alternative proposals from the base proposal, and major subcontractors that will be used and a completion date.

The general conditions for building construction consist of legal and other regulations, for the owner's protection, that the contractor must comply with during the course of the project. Special requirements supplement the general conditions by including specific information pertaining to this project. They will also include directions on use of water and electricity on the site, location of the contractors' field offices and storage areas, contractors' employee parking areas, and security measures.

The planning diagram prepared for the final design phase is shown in Figure 8.9. The manual calculations made from this diagram indicate a mid-November, 1996 completion date, which satisfies the FGD objective.

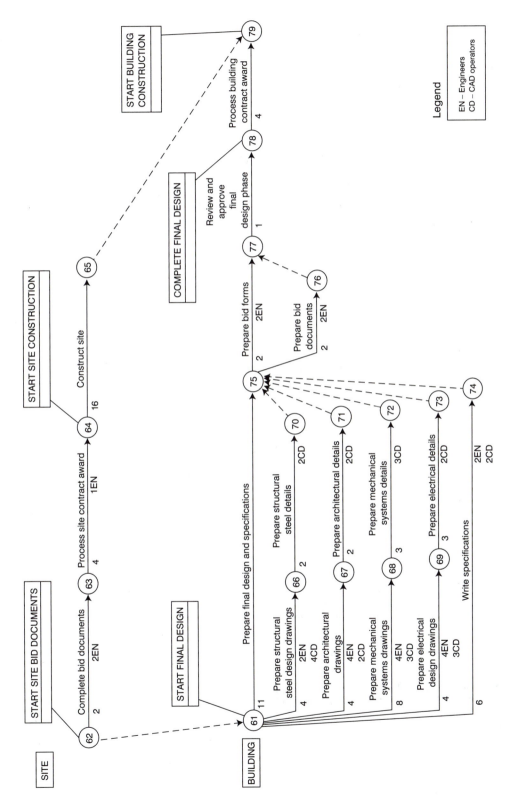

FIGURE 8.9 Planning diagram/final design phase for the FGD facility project

187

(a)

```
DESIGN CONCEPTS SCHEDULE                    FGD FACILITY PROJECT                        NASHVILLE, TENNESSEE
     REPORT TYPE :MILESTONE                                         PRINTING SEQUENCE :Earliest Activities First
                                                                   SELECTION CRITERIA :MILESTONE
     PLAN I.D.   :FGDDC    VERSION  7                               TIME NOW DATE      :16/JAN/96
===============================================================================================================
          MILESTONE                          EARLIEST EVENT          LATEST EVENT
          DESCRIPTION                         TIME                    TIME
===============================================================================================================
  1-   1 (START PROJECT) START SITE REQUIRE   16/JAN/96               16/JAN/96
 20-  20 START USERS (OWNERS) REQUIREMENTS    30/JAN/96               27/FEB/96
 26-  26 COMPLETE USERS REQUIREMENTS           9/APR/96               16/APR/96
 27-  27 COMPLETE DESIGN CONCEPTS PHASE       16/APR/96               23/APR/96
---------------------------------------------------------------------------------------------------------------
 15-  15 PURCHASE SITE                        23/APR/96               23/APR/96
===============================================================================================================
```

(b)

```
PRELIMINARY DESIGN SCHEDULE                 FGD FACILITY PROJECT                        NASHVILLE, TENNESSEE
     REPORT TYPE :MILESTONE                                         PRINTING SEQUENCE :Earliest Activities First
                                                                   SELECTION CRITERIA :MILESTONE
     PLAN I.D.   :FGDPD    VERSION  4                               TIME NOW DATE      :17/APR/96
===============================================================================================================
          MILESTONE                          EARLIEST EVENT          LATEST EVENT
          DESCRIPTION                         TIME                    TIME
===============================================================================================================
 40-  40 START SITE DESIGN                    17/APR/96               17/APR/96
 50-  50 START DETAILED CONSTRUCTION STUDY    29/MAY/96               29/MAY/96
 49-  49 COMPLETE SITE DESIGN                  5/JUN/96               17/JUL/96
 60-  60 COMPLETE DETAILED CONSTRUCTION STUDY 17/JUL/96               17/JUL/96
---------------------------------------------------------------------------------------------------------------
===============================================================================================================
```

(c)

```
FINAL DESIGN PHASE                          FGD FACILITY PROJECT                        NASHVILLE, TENNESSEE
     REPORT TYPE :MILESTONE                                         PRINTING SEQUENCE :Earliest Activities First
                                                                   SELECTION CRITERIA :MILESTONE
     PLAN I.D.   :FGDFD    VERSION  4                               TIME NOW DATE      :17/JUL/96
===============================================================================================================
          MILESTONE                          EARLIEST EVENT          LATEST EVENT
          DESCRIPTION                         TIME                    TIME
===============================================================================================================
 62-  62 START SITE BID DOCUMENTS            17/JUL/96               17/JUL/96
 61-  61 START FINAL DESIGN PHASE            17/JUL/96               14/AUG/96
 64-  64 START SITE CONSTRUCTION             28/AUG/96               28/AUG/96
 78-  78 COMPLETE FINAL DESIGN PHASE         23/OCT/96               20/NOV/96
---------------------------------------------------------------------------------------------------------------
 79-  79 START BUILDING CONSTRUCTION         18/DEC/96               18/DEC/96
 65-  65 COMPLETE SITE CONSTRUCTION          18/DEC/96               18/DEC/96
===============================================================================================================
```

FIGURE 8.10   Milestone schedule for (a) design concepts; (b) preliminary design; and (c) final design phases for the FGD facility project

PERSONNEL ALLOCATION (INCLUDING CADD OPERATORS) FOR THE FGD PROJECT

As engineering and CADD systems personnel are most commonly used in this type of design effort, GAI will first examine this demand over the entire course of the project. Using the durations and the allowances of engineering and CADD personnel for each activity shown on the planning diagrams of the three phases, GAI can generate computerized timing schedules and computerized personnel load charts. The timing schedules help in ascertaining the likelihood of completing the engineering project. GAI plans to submit with the proposal the computerized milestone schedules for each phase (Figure 8.10) to confirm that the plan will meet the objectives. GAI will not submit the detailed timing schedules nor the engineer and CADD operator load charts used in developing the pricing schedule for the proposal. These are essentially worksheets, needing further refinements, and could be misleading if subject to scrutiny by anyone other than GAI. Furthermore, they are just part of the pricing and a great deal of information is yet to be accumulated before completing the final proposal.

From the flow curves generated, the total CADD operator personnel requirements are as follows:

Design Phase	CADD-Days	CADD-Hours (CADD-Days × 8)	Total $ (CADD-Hours × 50)
Design concepts	282	2256	112,800
Preliminary design	382	3056	152,800
Final design	463	3704	185,200
Totals	1127	9016	$450,800

The computer-generated CAD-days needed to complete the final design phase are shown in Figure 8.11.

For a CADD operator at a work station, GAI charges $50 an hour. These charges include equipment depreciation, maintenance allowances, certain overhead, and direct supervision costs.

In a similar manner, GAI calculated the Engineer load and charges to be included in the proposal:

$$1368 \text{ Engineer-days} \times 8 \text{ hours/day} \times \$52.50/\text{hour} = \$574,560$$

The total pricing breakdown accommodating all of the items that GAI will be including in their proposal would be as follows:

CADD	$450,000
engineers (all disciplines)	574,000
supplemental engineering support	150,000
project manager (including secretary)	123,000
administrative support	60,000
materials	50,000
subtotal	$1,407,000
general overhead (10%)	140,000
subtotal	$1,547,000
profit (10%)	160,000
total	$1,707,000

GAI would show just the total price in the proposal unless FGD had specifically requested a breakdown. This price does include a certain contingency to allow for any uncertainties that FGD would not cover.

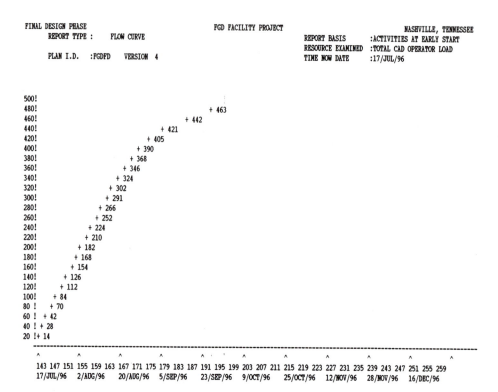

FIGURE 8.11   Total CAD operator load (CAD-days) for the final design phase of the FGD project

GAI believes that the allocation of the engineering and CADD personnel can be improved once the project is under way. If the firm succeeds in effecting these improvements, it will be able to increase the profit margin. These improvements depend on upgrading the CADD operators' role in the design process.

Role of CADD Personnel

Since the onset of the company's operations, GAI has recognized the importance of using CADD in the design process. At GAI, CADD work stations have nearly replaced the traditional drafting boards.

Using computer-aided drafting in preparing drawings for design and engineering projects has reduced time and costs a number of ways:

- Design alternatives can be produced quickly. The basic design can be readily changed to show any number of optional schemes.
- Design modifications can be incorporated quickly.
- Each design alternative or version can be filed and recovered with ease.
- Printouts of each design drawing in varying stages can be produced for review while continuing to work on the drawings.

- An important time-reducing feature in producing drawings is that more than one CADD person can work on the same drawing at the same time.

GAI found that most designers and engineers, who had some CADD training and had been assigned to CADD work stations, were uncomfortable in this type of environment. They found it awkward combining their roles of design and coordinating with other designers, engineers, and CADD personnel while at a CADD work station. This use of equipment is inefficient, since designers and engineers spend much of their time creating, calculating, coordinating, and checking drawings of CADD operators under their supervision. The CADD equipment should be kept producing drawings, whether they are originals or revisions.

While the long-term objectives call for a greatly increased use of CADD equipment by designers and engineers, the short-term solution for GAI is to establish a separate CADD group. This section, headed by a director on the same level as the engineering and design disciplines, is made up of people very proficient in using CADD for drafting work. The group will continually be expanded with qualified CADD personnel with design training and experience performing selected design work. To accomplish this, GAI will

- encourage the present CADD operators to acquire design skills through further education and training, and gradually assign them detailed design work
- recruit qualified CADD operators with drafting skills and training and with some basic technical design knowledge and education

CONCLUSION

Preparing an engineering proposal for designing a building facility parallels the procedure followed in carrying out a project. It follows these steps:

1. Once the objectives are set, the design is planned in phases: design concepts, preliminary design, and final design. The backbone of the plan is the WBS, which not only defines the work that must be done, but can also depict the personnel assigned to assist the project leader.
2. A computerized milestone schedule is generated to determine the plausibility of the objectives.
3. Personnel required to complete each work activity are scheduled.

Generating accumulated work load flow curves are used for (1) securing personnel totals for pricing purposes when preparing the proposal and (2) status reporting on the total number of personnel used to date. Histograms are generated to show peak loadings and determine whether they can be handled in-house or whether they require outsourcing (and, if so, at what additional cost).

An engineering proposal that summarizes the project plan and demonstrates schedule credibility with a dependable personnel strategy will receive favorable consideration. It essentially unfolds into the project plan.

Exercises

1. Using the planning diagrams in this chapter, prepare the computerized schedules for the following:
   a. Design concepts phase
   b. Preliminary design phase
   c. Final design phase
2. FGD set up the following milestones for its facility:

Event	FGD Milestone	GAI Milestone
Start project	1/16/96	_____
Complete site engineering and design	6/15/96	_____
Groundbreaking ceremonies (start site work construction)	9/01/96	_____
Complete building design	11/15/96	_____
Start building construction	12/15/96	_____

   a. Using the data from Figure 8.9, complete the GAI Milestone column.
   b. Do the FGD and GAI milestone dates compare favorably? (Note: Favorable milestone dates would refer to the GAI milestones at least one week—five days—earlier to allow for possible drifting of any pertinent critical activities. To determine the course of action of the unfavorable activities, if any, complete the following table:

Event	Comparison of Milestones GAI Earlier/ (Later) Than FGD (Days)	Action
Start project	_____	_____
Complete site engineering and design	_____	_____
(start site work construction)	_____	_____
Complete building design	_____	_____
Start building construction	_____	_____

(Suggestion: Consider reducing the duration of selected activities leading to the unfavorable milestone. If they are required, prepare a new schedule tabulation to see the effect on other activities.)

3. The GAI proposal will not include any personnel loading charts; however, they have an important role in planning the personnel supply for the project when it begins. GAI is interested in knowing the following information:

a. The maximum number of CADD operators required for each phase.

b. The length of these peak periods.

c. The percent of the total CADD staff required for this period.

Planning and Scheduling a Building Project

Possessing a thorough plan is among the foremost rules for proceeding with a construction project. Preparing a plan starts once the contractors receive the notice to proceed and the main timing and cost objectives, including any interim objectives (or milestones). After they have initiated a series of exercises to determine what work has to be done and developed the work breakdown structure (WBS), they prepare the project planning diagram. Time estimates and resources (including costs) data are then allocated and added to the diagram to complete the plan. At this stage, the planning diagram becomes a working document that helps in executing initial project strategies. The planning diagram allows scheduling the project to begin—it acts as a guidebook for the contractors to use in managing the project.

The scheduling process consists of the exercises to determine when the work has to be done and the distribution of personnel/labor and costs. The computer is essential in calculating the schedules for the timing and costs of the project. Reactions to changes in the field must be fast, and contractors want quick resolutions to problems that will arise during the project. When changes to the project occur, contractors may need to analyze several situations and evaluate alternative plans. A good computer program will examine what-if situations quickly and easily. Proper software must be easy, clear, and not time-consuming for initial training and effectiveness in helping to manage the project.

Time and cost management are the keys to launching and competently supervising building construction projects. Also significant to the success of the project are the personnel who will be supervising schedules and costs.

CONSTRUCTION MANAGEMENT

Construction management (CM) is sometimes used synonymously with project management in construction projects. While both concern managing projects, CM entails much broader responsibilities. One of the added responsibilities of CM is handling and coordinating all of the construction subcontractors on the project for the owner. Another significant CM responsibility includes administering the subcontractor contracts; to be done efficiently, this function necessarily encompasses project management functions.

The owner can discharge the duties of CM, designate them to an architectural/engineering firm (A&E) or assign them to the general contractor who may specialize in CM along with standard general contracting work. The general contractor typically acts as CM, with the A&E providing the building designs and specifications and the CM handling all affairs of the building construction project. Construction management is not the same job as the managing arrangement used in the Christopher Design/Build Project described in the earlier portions of this book. In CM, one organization is responsible for the entire project.

The main responsibilities of CM are as follows:

- Under the CM approach, three or four contractors will officially bid on each major section of the work defined in the specifications. With a general contract, there may be as many bidders, but the owner may not be familiar with them. This may cause possible doubt in some cases of whether the project will be using the best available contractors.
- Cost management is more effective cost because of better control of the project. CM provides a better understanding of all participants' roles, reducing uncertainties that are the main cause for high bids.
- The CM concept inherently encourages disciplined planning, scheduling, and control for timing and cost. As a result, contractors need to provide more detailed information in their bids. Upon a successful proposal they add this detail as part of their plan of action.
- CM relieves the architect/engineer of the riskier and less profitable functions—estimating, project quality control and inspection, construction methods, and field supervision inspection. This arrangement works well, since the CM usually has had more experience in these areas.

For a more effective CM arrangement, contractors do most of the construction field work while the construction manager's role in the field is minor, concentrating more on administrative duties. CM tries to create an environment to produce project coordination through communication. At weekly meetings, representatives of the owner, architect/engineer, and construction manager review firsthand the job's progress, the finest example of communicating. The construction manager has companion weekly meetings with the contractors to coordinate the work along with such subjects as ensuring worker satisfaction, safety, and productivity.

Construction management involves many functions, including

1. acting as liaison among all participants of the project
2. preparing the construction plan
3. generating the master schedule based upon the schedules of all of the subcontractors
4. monitoring the project performance
5. administering the performance payments
6. forecasting cash flow
7. overseeing quality
8. maintaining all schedules

BACKGROUND: SOUTHFORK CONSTRUCTION PROJECT

The City of Southfork plans to construct a building addition to the library building at its civic center. The addition consists of a 50,000-square-foot structure with exterior and interior characteristics similar to those of the main building. Southfork has already retained S&E Associates, an architectural and engineering firm, to prepare the construction drawings, specifications, and bid documents. The city awarded the construction manager contract to Lawrence Construction Company (LCC) for a not-to-exceed price of $4,450,000. Besides the construction manager fees, the contract provides funds to furnish labor, material, and equipment to complete the project. An exception to this contract is a provision that the City of Southfork will furnish several major long-term delivery building equipment items. This type of arrangement is necessary for timely shipments of any critical items to avoid jeopardizing the construction schedule. A separate responsibility of this nature may be awkward—how to accept shipment at the job site, problems associated with defects, potential uncoordinated responsibilities—and all of these require special details in the specifications. S&E, which will be designing the long-term equipment, will also assume responsibility for preparing the purchase order documents, processing the order, and coordinating the fabrication process and shipment to the job site. To complete the long-term equipment installation, S&E will include in the specifications that the successful contractor (LCC) will be responsible for acceptance of this equipment and be accountable for proper installation and tryout.

Terms of the contract provide for construction to start by March 4, 1996 and finish by February 10, 1997, when the City of Southfork will begin occupancy.

PLANNING THE PROJECT

Under the terms of the contract, LCC will provide a detailed plan of action to satisfy the City of Southfork that the operational plan meets the city's objectives. To administer the project, LCC has divided it into three phases: planning, scheduling, and controlling.

The planning phase has three main steps:

1. Establish project objectives.
2. Develop the project plan.
3. Prepare the project planning diagram.

The Project Objectives

After reviewing proposals on the major work divisions to be done, LCC selected the following contractors:

Contractor	Major Work Responsibilities
Nardan Constructors	excavation
	concrete
Kanes Steel	structural steel
	roofing and siding
Beede Builders	building exterior
	building interiors
	cleanup and approval

At the initial organizational meeting, LCC reviews the main timing and cost objectives set by the City of Southfork:

- Start project   March 4, 1996
- Complete project; building ready
  for occupancy   February 10, 1997
- Authorized budget   Not to exceed: $4,450,000

Using the starting and completion dates as the base, LCC and the designated contractors set up the interim objectives or milestones. Milestones are defined as the significant events to be achieved before meeting the main objectives. The milestones set with the responsible contractors are as follows:

Milestone	Contractor	Start	Complete
Start project	LCC	3/4/96	—
Excavation	Nardan	3/4/96	5/20/96
Concrete	Nardan	3/4/96	6/17/96
Structural steel	Kanes	3/4/96	7/29/96
Roofing	Kanes	6/17/96	9/7/96
Building exterior	Beede	3/4/96	10/4/96
Building interiors	Beede	7/29/96	12/16/96
Cleanup and approval	Beede	12/16/96	2/10/97
Complete project	LCC	—	2/10/97

FIGURE 9.1 Milestone bar chart for the Southfork Building Project

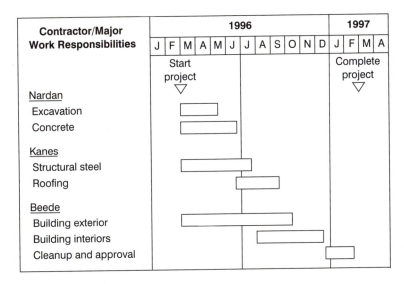

Figure 9.1, which graphically depicts the milestones, is distributed to the contractors. They will use this as a guide to prepare their plan of action and schedule.

The Plan of Action

The plan of action is made up of the activities needed to meet the project's objectives and milestones. When all the contractors make up a list of the activities needed to complete their work, the construction manager will organize them in a logical sequence to prepare a project planning diagram. The completed diagram will supply the necessary input data to produce the computerized time and cost schedules. Added to the time and cost information will be resource planning data, also identified on the diagram, for generating the construction labor/personnel loadings.

The steps leading to construction of the project planning diagram using the project management approach are as follows:

1. List the important activities.
2. Develop the work breakdown structure (WBS).
3. List the personnel/labor required to complete each activity.
4. Identify personnel/labor costs, material and other costs to complete each activity.

Work Breakdown Structure. Figure 9.2 shows a completed WBS for the Southfork Building Project. The WBS is a format that the contractors use to identify and organize the activities necessary to complete a project. It is an excellent way to show the role of the significant activities. It also helps point out any omissions or duplications before they cause schedule inaccuracies. Preparing the WBS for many projects is difficult and tedious. Nonetheless, when completed, it proves most rewarding as the process compels the participants to think through their project.

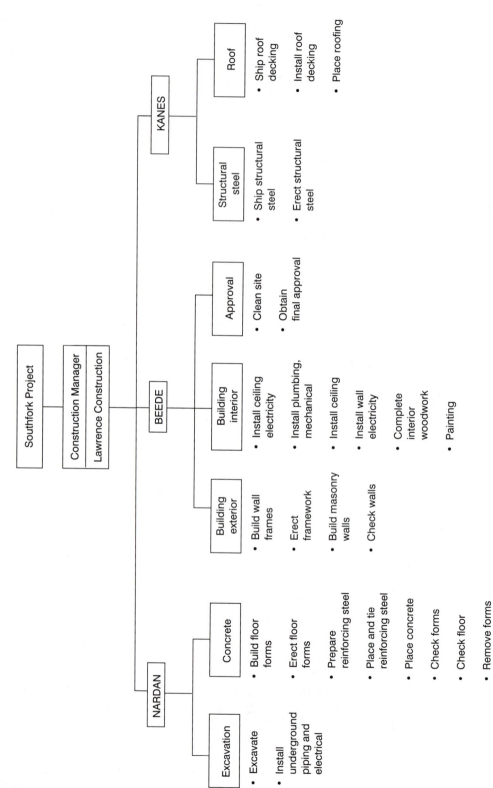

FIGURE 9.2 Work breakdown structure for the Southfork Building Project

There are varying styles of WBS formats which will show the responsibilities and/or resources with each activity. This may be a more desirable option for reporting purposes. Those not familiar with network diagramming might prefer to review the organizational charting used in the WBS format. They could be confused with the appearance of a project planning diagram and uncertain of its purpose. Regardless, identifying all the data on the project planning diagram is important for preparation for computer analysis.

The Project Planning Diagram.   Once the WBS has been completed to the mutual satisfaction of the construction manager and the contractors, they investigate the relationships of the activities it shows. Each contractor examines its own activities, determines how they are connected, and then prepares a preliminary diagram (or subdiagram) as shown in Figure 9.3.

Once each contractor has completed its respective subdiagram and reviewed it with the construction manager, there will be a joint meeting among all of the contractors to investigate the interrelationships among all of their work activities. Figure 9.4 shows the completed subdiagram of this effort. Preparing subdiagrams will simplify constructing the final project plan.

Constructing the final project plan diagram, using the stated rules of arrow diagramming, gives an accurate portrayal of the project. Added to this diagram will be the costs and resources required for each individual activity. The construction manager will test the validity of the costs, resources, and timing estimates while the contractors are completing the final project plan. Figure 9.5 shows the completed project planning diagram for the Southfork Building Project with durations, costs, and resources of each activity.

After reviews and approvals of the project planning diagram, the data is now available to develop the computerized schedule. While most software programs can generate schedule reports showing the timing and required resources using data entered from the planning diagram, very few, if any, can effectively draw a thorough, logical plan structured in the desired WBS form. Drawing it manually or with computer-aided design (CAD) provides a clear, concise, workable plan. The quality of computer graphics used to display a project diagram is improving; eventually, it will likely meet the required standards.

THE PROJECT SCHEDULE

The scheduling phase concerns when the work will be done. The scheduling techniques used will give LCC and the contractors worthwhile timing information by

- establishing a supportable project duration for the plan
- identifying the critical activities that make up the critical path through the project
- identifying the activities where schedules can be altered without affecting the project duration

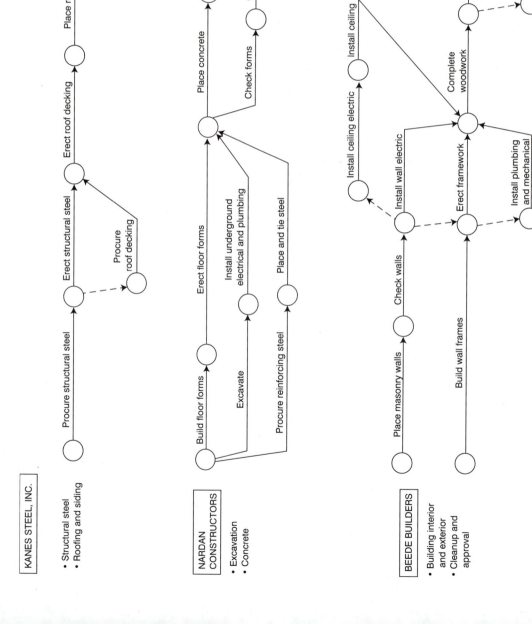

FIGURE 9.3 Connections among each contractors' activities on the Southfork Building Project

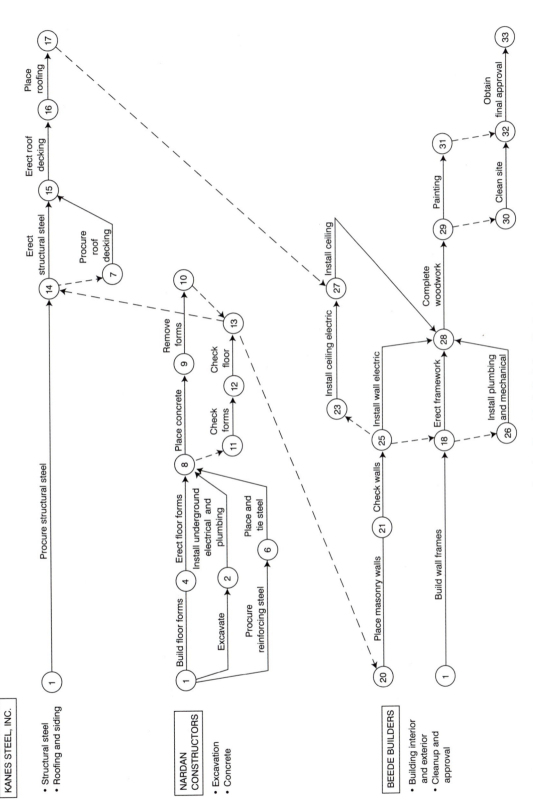

FIGURE 9.4 Interrelationships among contractors for the Southfork Building Project

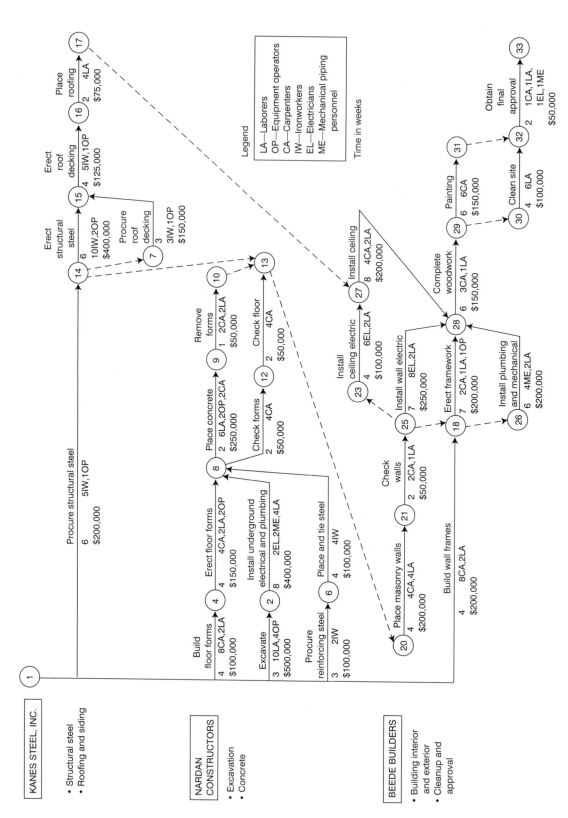

FIGURE 9.5 Completed planning diagram for the Southfork Building Project

Project Time Estimates

Time estimates are the basis of project scheduling. To eliminate any potential inconsistencies, LCC and the contractors jointly prepared time estimates needed to complete the activities in the project. Their knowledge of their own work is the key to achieving acceptable estimates. They had to ensure that contingency factors that contractors would want to include in their estimates for unforeseen conditions did not become excessive. The time estimating process can also become a series of compromises among the contractors. The "give and take" discussions continue until all parties are in accord. The acceptable time estimates finally decided upon were placed below each activity arrow on the approved planning diagram.

Manual Timing

Before using computer scheduling programs, LCC will use certain parts of the manual scheduling method for several quick timing checks. This is a quick way to determine if the project duration (as defined by the objectives) conforms with the timing data shown on the project planning diagram. Calculating the earliest start times of each project activity will give an immediate estimate of project duration. It will also acquaint those working on the project with the analysis and logic used in scheduling projects by this technique.

Figure 9.6 illustrates these efforts. It displays the estimates and earliest start calculations for the activities on the project planning diagram that produce the acceptable project duration date. Also shown is the project duration, which is the earliest start time of the last activity of the project. It is 47 weeks from the March 4, 1996 start date (using a calendar or the date calculator, that would be February 10, 1997). Manual scheduling may be used when (1) computer time is restricted, (2) using a computer is not feasible for scheduling small projects, or (3) making relatively small adjustments to the original plan. It is important to understand both manual schedule calculations and computerized scheduling. Use whatever method requires the least time and effort to validate the main objectives before starting the larger scheduling task. If the calculated project duration is not acceptable, then keep adjusting estimates and the logic until the timing objectives of the project are satisfied.

Using the Computer for Scheduling

Scheduling for a project of this nature and size needs be done with appropriate project management software. This project is too large in scope to manually calculate the time, cost, and resource schedules. Countless computer programs are capable of producing the needed computerized scheduling reports. As LCC and the contractors have no personnel familiar with advanced project management software, they elect to use a program that is quick and easy to learn and use. (Many consulting firms specialize in applying project management software to

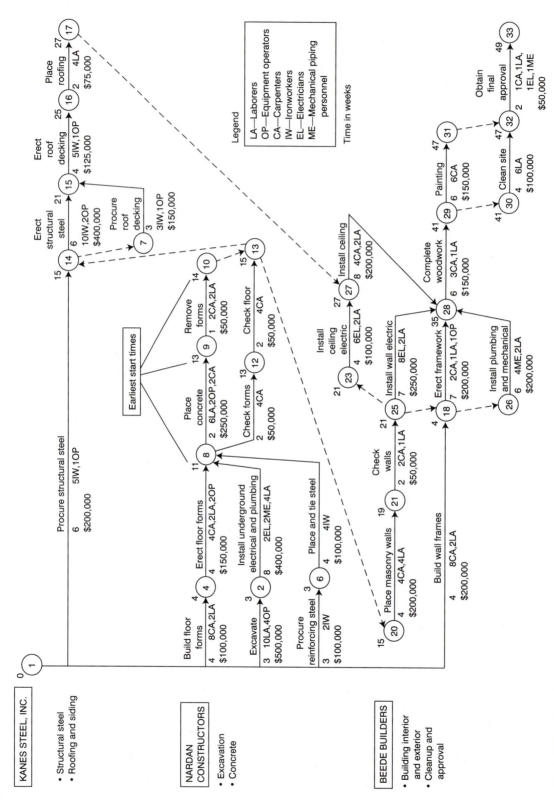

FIGURE 9.6 Earliest start times for the Southfork Building Project

clients' projects. It is a matter of cost effectiveness—spending the time and effort on this project training your own personnel for future applications may be more practical.)

Computers are a necessary part of construction operations. In addition to scheduling, the construction manager and contractors will use it frequently for analysis, especially for what-if situations, to improve the overall project. As the Southfork Building Project is a typical dynamic project, changes are to be expected. Project management software must provide proper assistance for handling project concerns resulting from inevitable changes as the project is progressing. While attractive graphics on the screen and on generated reports are desirable for reporting the status of a project, this is not necessarily a priority for the contractor out in the field who needs a computer program that provides excellent analytical assistance.

The software packages available for personal computers (PC) have improved appreciably during recent years. PCs have become so powerful that they have essentially replaced miniframe and mainframe computer systems for project management purposes. While using the personal computer for schedule calculations has many advantages over manual methods, the greatest benefit is speed. Computer-generated new and revised bar charts are much faster than making manual charts. Analyzing what-if situations without disturbing the base schedule is a decided advantage. The ability to compare personnel required with available skilled personnel becomes a valuable asset with the use of the personal computer.

Software suppliers are constantly upgrading and improving their products. The software package chosen for this project is more fundamental than many now available. It is especially adaptable to meet the needs of the projects used in this book: It is easy to learn, easy to use, and compatible with the arrow diagramming and the precedence diagramming techniques. The software can provide these reports:

- Tabulated schedules for the project activities (construction schedules) showing the earliest and latest start dates, earliest and latest finish dates, duration, and float times. Through a "filter" sort, each contractor can be limited to its portion of the schedule.
- Bar chart schedules that graphically portray the information shown in the tabulated schedule. The bar chart schedules supplement the tabulated schedules, and many contractors prefer this graphic display. Bar charts are especially useful for analyses.
- Milestone reports summarize the plan and schedule and are used when reviewing all of the project. Problems are detected more easily when using milestone reports.
- Project control reports for distribution to suppliers and contractors.
- Cost histograms and cash flow curves will plan and analyze the cost schedules. They will be used to compare with the availability of funds allotted for this project.
- Resource histograms and summary resource reports will compare the required number of critical skilled tradesmen with the available number set aside for this project.

Construction Schedules

LCC will be reviewing and updating the construction schedules continually as part of the routine procedures of planning daily operations and evaluating performance. Schedules are useful for planning the future functions of the contractors. Figure 9.7 shows the schedule, sorted by the individual contractors on the job, for the total project. Printing schedules of individual contractors' timing is common. All of the essential timing details associated with the work are successively positioned chronologically. LCC can scan the list and review the current planned work schedule of any of the contractors at the start of the day of any given date. Any scheduled date departing from the plan can be checked immediately against its timing flexibility (float time) and for any possible effect on the project's duration.

Another important construction schedule report is the critical activities listing that highlights the "0" float and the low float activities. At the onset of the project, LCC can immediately observe the impact of individual activities on its

```
PROJECT MANAGEMENT                        SOUTHFORK BUILDING PROJECT                          SOUTHFORK, PENNSYLVANIA
        REPORT TYPE :STANDARD LISTING                                        PRINTING SEQUENCE  :Node/Activity Sequence
                                                                             SELECTION CRITERIA :ALL
        PLAN I.D.  :SBP      VERSION  16                                      TIME NOW DATE      : 4/MAR/96
```

ACTIVITY DESCRIPTION	EARLIEST START	EARLIEST FINISH	LATEST START	LATEST FINISH	DURATION	FLOAT
1- 18 BE BUILD WALL FRAMES	4/MAR/96	29/MAR/96	15/AUG/96	11/SEP/96	20	118
18- 28 BE ERECT FRAMEWORK	1/APR/96	17/MAY/96	12/SEP/96	30/OCT/96	35	118
20- 21 BE BUILD MASONRY WALLS	17/JUN/96	12/JUL/96	17/JUN/96	12/JUL/96	20	0 *
21- 23 BE CHECK WALLS	15/JUL/96	26/JUL/96	15/JUL/96	26/JUL/96	10	0 *
26- 28 BE INSTALL PLUMBING,MECHANICAL	29/JUL/96	6/SEP/96	19/SEP/96	30/OCT/96	30	38
23- 28 BE INSTALL WALL ELECTRICAL	29/JUL/96	13/SEP/96	12/SEP/96	30/OCT/96	35	33
25- 27 BE INSTALL CEILING ELECTRICAL	29/JUL/96	4/SEP/96	29/JUL/96	4/SEP/96	28	0 *
27- 28 BE INSTALL CEILING	5/SEP/96	30/OCT/96	5/SEP/96	30/OCT/96	40	0 *
28- 29 BE COMPLETE INTERIOR WOODWORK	31/OCT/96	11/DEC/96	31/OCT/96	11/DEC/96	30	0 *
29- 31 BE PAINT INTERIOR,EXTERIOR	12/DEC/96	22/JAN/97	12/DEC/96	22/JAN/97	30	0 *
30- 32 BE CLEANUP SITE	12/DEC/96	20/JAN/97	16/DEC/96	22/JAN/97	28	2
32- 33 BE OBTAIN FINAL APPROVAL	23/JAN/97	5/FEB/97	23/JAN/97	5/FEB/97	10	0 *
1- 14 KA PROCURE STRUCTURAL STEEL	4/MAR/96	12/APR/96	28/MAY/96	8/JUL/96	30	61
14- 15 KA ERECT STRUCTURAL STEEL	17/JUN/96	12/JUL/96	9/JUL/96	5/AUG/96	20	16
7- 15 KA PROCURE ROOF DECKING	17/JUN/96	5/JUL/96	16/JUL/96	5/AUG/96	15	21
15- 16 KA ERECT ROOF DECKING	15/JUL/96	9/AUG/96	6/AUG/96	2/SEP/96	20	16
16- 17 KA PLACE ROOFING	12/AUG/96	13/AUG/96	3/SEP/96	4/SEP/96	2	16
1- 2 NA EXCAVATE	4/MAR/96	22/MAR/96	4/MAR/96	22/MAR/96	15	0 *
1- 6 NA PROCURE REINFORCING STEEL	4/MAR/96	22/MAR/96	1/APR/96	19/APR/96	15	20
1- 4 NA BUILD FLOOR FORMS	4/MAR/96	29/MAR/96	25/MAR/96	19/APR/96	20	15
2- 8 NA INSTALL UNDERGROUND PIPING & ELEC	25/MAR/96	17/MAY/96	25/MAR/96	17/MAY/96	40	0 *
6- 8 NA PLACE & TIE REINFORCING STEEL	25/MAR/96	19/APR/96	22/APR/96	17/MAY/96	20	20
4- 8 NA ERECT FLOOR FORMS	1/APR/96	26/APR/96	22/APR/96	17/MAY/96	20	15
8- 9 NA PLACE CONCRETE	20/MAY/96	31/MAY/96	27/MAY/96	7/JUN/96	10	5
11- 12 NA CHECK FORMS	20/MAY/96	31/MAY/96	20/MAY/96	31/MAY/96	10	0 *
12- 13 NA CHECK FLOOR	3/JUN/96	14/JUN/96	3/JUN/96	14/JUN/96	10	0 *
9- 10 NA REMOVE FORMS	3/JUN/96	7/JUN/96	10/JUN/96	14/JUN/96	5	5

FIGURE 9.7   Construction schedule for the Southfork Building Project

duration. While the contractors now are aware that all their activities are critical, LCC will point out specific items for special treatment. Figure 9.8 shows this type of report. A good example of its effectiveness is the handling of the *Excavate* activity. This highlighted "0" float item must start on March 4, 1996 to avoid any potential schedule problems. Nardan, who has responsibility for this item, will mobilize ahead of the March 4 date in preparation of the start. From this same report Nardan can note that the *Excavate* activity precedes a series of successive activities with 0 float. Nardan's personnel on this project will closely watch to make sure the performance of their work stays on schedule from the period of March 4 to June 14. The work of other contractors is dependent on Nardan keeping on schedule.

Bar Chart Schedules.   For this project, LCC uses two types of bar charts. The contractors will use the daily bar chart for ongoing operations and LCC will use the summary period bar chart for analysis and reporting purposes.

```
PROJECT MANAGEMENT                        SOUTHFORK BUILDING PROJECT                      SOUTHFORK, PENNSYLVANIA
       REPORT TYPE :STANDARD LISTING                                   PRINTING SEQUENCE  :Most Critical Activities First
                                                                       SELECTION CRITERIA :ALL
       PLAN I.D.  :SBP      VERSION  16                                TIME NOW DATE      : 4/MAR/96
=================================================================================================================
       ACTIVITY DESCRIPTION                  EARLIEST     EARLIEST     LATEST       LATEST      DURATION  FLOAT
                                             START        FINISH       START        FINISH
=================================================================================================================
  1-   2 NA EXCAVATE                         4/MAR/96     22/MAR/96    4/MAR/96     22/MAR/96      15        0 *
  2-   8 NA INSTALL UNDERGROUND PIPING & ELEC 25/MAR/96   17/MAY/96    25/MAR/96    17/MAY/96      40        0 *
 11-  12 NA CHECK FORMS                       20/MAY/96   31/MAY/96    20/MAY/96    31/MAY/96      10        0 *
 12-  13 NA CHECK FLOOR                       3/JUN/96    14/JUN/96    3/JUN/96     14/JUN/96      10        0 *
-----------------------------------------------------------------------------------------------------------------
 20-  21 BE BUILD MASONRY WALLS              17/JUN/96    12/JUL/96    17/JUN/96    12/JUL/96      20        0 *
 21-  23 BE CHECK WALLS                      15/JUL/96    26/JUL/96    15/JUL/96    26/JUL/96      10        0 *
 25-  27 BE INSTALL CEILING ELECTRICAL       29/JUL/96    4/SEP/96     29/JUL/96    4/SEP/96       28        0 *
 27-  28 BE INSTALL CEILING                  5/SEP/96     30/OCT/96    5/SEP/96     30/OCT/96      40        0 *
-----------------------------------------------------------------------------------------------------------------
 28-  29 BE COMPLETE INTERIOR WOODWORK       31/OCT/96    11/DEC/96    31/OCT/96    11/DEC/96      30        0 *
 29-  31 BE PAINT INTERIOR,EXTERIOR          12/DEC/96    22/JAN/97    12/DEC/96    22/JAN/97      30        0 *
 32-  33 BE OBTAIN FINAL APPROVAL            23/JAN/97    5/FEB/97     23/JAN/97    5/FEB/97       10        0 *
 30-  32 BE CLEANUP SITE                     12/DEC/96    20/JAN/97    16/DEC/96    22/JAN/97      28        2
-----------------------------------------------------------------------------------------------------------------
  8-   9 NA PLACE CONCRETE                   20/MAY/96    31/MAY/96    27/MAY/96    7/JUN/96       10        5
  9-  10 NA REMOVE FORMS                     3/JUN/96     7/JUN/96     10/JUN/96    14/JUN/96      5         5
  1-   4 NA BUILD FLOOR FORMS                4/MAR/96     29/MAR/96    25/MAR/96    19/APR/96      20        15
  4-   8 NA ERECT FLOOR FORMS                1/APR/96     26/APR/96    22/APR/96    17/MAY/96      20        15
-----------------------------------------------------------------------------------------------------------------
 14-  15 KA ERECT STRUCTURAL STEEL           17/JUN/96    12/JUL/96    9/JUL/96     5/AUG/96       20        16
 15-  16 KA ERECT ROOF DECKING               15/JUL/96    9/AUG/96     6/AUG/96     2/SEP/96       20        16
 16-  17 KA PLACE ROOFING                    12/AUG/96    13/AUG/96    3/SEP/96     4/SEP/96       2         16
  1-   6 NA PROCURE REINFORCING STEEL        4/MAR/96     22/MAR/96    1/APR/96     19/APR/96      15        20
-----------------------------------------------------------------------------------------------------------------
  6-   8 NA PLACE & TIE REINFORCING STEEL    25/MAR/96    19/APR/96    22/APR/96    17/MAY/96      20        20
  7-  15 NA PROCURE ROOF DECKING             17/JUN/96    5/JUL/96     16/JUL/96    5/AUG/96       15        21
 23-  28 BE INSTALL WALL ELECTRICAL          29/JUL/96    13/SEP/96    12/SEP/96    30/OCT/96      35        33
 26-  28 BE INSTALL PLUMBING,MECHANICAL      29/JUL/96    6/SEP/96     19/SEP/96    30/OCT/96      30        38
-----------------------------------------------------------------------------------------------------------------
  1-  14 KA PROCURE STRUCTURAL STEEL         4/MAR/96     12/APR/96    28/MAY/96    8/JUL/96       30        61
  1-  18 BE BUILD WALL FRAMES                4/MAR/96     29/MAR/96    15/AUG/96    11/SEP/96      20        118
 18-  28 BE ERECT FRAMEWORK                  1/APR/96     17/MAY/96    12/SEP/96    30/OCT/96      35        118
=================================================================================================================
```

FIGURE 9.8   Critical activity listings for the Southfork Building Project

```
PROJECT MANAGEMENT                    SOUTHFORK BUILDING PROJECT                  SOUTHFORK, PENNSYLVANIA
    REPORT TYPE :PERIOD BARCHART                                  PRINTING SEQUENCE  :Node/Activity Sequence
                                                                  SELECTION CRITERIA :ALL
       PLAN I.D.  :SBP     VERSION 12                             TIME NOW DATE      : 4/MAR/96
==================================================1996===============================================================
    PERIOD COMMENCING DATE            !4   !11  !18  !25  !1   !8   !15  !22  !29  !6   !13  !20  !27  !3   !
    MONTH                             !MAR !    !    !    !APR !    !    !    !    !MAY !    !    !    !JUN  !
    PERIOD COMMENCING TIME UNIT       !2   !7   !12  !17  !22  !27  !32  !37  !42  !47  !52  !57  !62  !67  !
======================================================================================================================
    1- 18 BE BUILD WALL FRAMES        !=====!=====!=====!=====!    !    !    !    !    !    !    !    !    !    !
   18- 28 BE ERECT FRAMEWORK          !    !    !    !    !=====!=====!=====!=====!=====!=====!=====!    !    !    !
    1- 14 KA PROCURE STRUCTURAL STEEL !=====!=====!=====!=====!=====!=====!    !    !    !    !    !    !    !    !
    1-  2 NA EXCAVATE                 !CCCCC!CCCCC!CCCCC!    !    !    !    !    !    !    !    !    !    !    !

    1-  4 NA BUILD FLOOR FORMS        !=====!=====!=====!=====!    !    !    !    !    !    !    !    !    !    !
    1-  6 NA PROCURE REINFORCING STEEL!=====!=====!=====!    !    !    !    !    !    !    !    !    !    !    !
    2-  8 NA INSTALL UNDERGROUND PIPING!   !    !    !CCCCC!CCCCC!CCCCC!CCCCC!CCCCC!CCCCC!CCCCC!CCCCC!    !    !    !
    6-  8 NA PLACE & TIE REINFORCING STE!   !    !    !=====!=====!=====!=====!    !    !    !    !    !    !    !

    4-  8 NA ERECT FLOOR FORMS        !    !    !    !    !=====!=====!=====!=====!    !    !    !    !    !    !
   11- 12 NA CHECK FORMS              !    !    !    !    !    !    !    !    !    !    !    !CCCCC!CCCCC!    !
    8-  9 NA PLACE CONCRETE           !    !    !    !    !    !    !    !    !    !    !    !=====!=====!    !
   12- 13 NA CHECK FLOOR              !    !    !    !    !    !    !    !    !    !    !    !    !    !CCCCC!>

    9- 10 NA REMOVE FORMS             !    !    !    !    !    !    !    !    !    !    !    !    !    !=====!>
======================================================================================================================
Barchart Key:- CCC :Critical Activities   === :Non Critical Activities   NNN :Activity with neg float
```

FIGURE 9.9 Bar chart of Southfork Building Project activities during the March 3 through June 10 period

The contractors' use of the daily bar chart (Figure 9.9) is an essential part of everyday operations. It will usually be placed on a wall of the office with progress and notations marked daily. The contractor can immediately observe the progress or lack of progress. This bar chart displays the required contractors' work from the period of March 2 to June 10. (Longer periods can be printed using the "landscape" print mode.) In addition, a sort can show the work of the individual contractors during that same period. This chart shows critical and non-critical activities. As LCC does not want the subcontractors to know what float is available, a feature will modify the report to eliminate that information. LCC requires the contractors to complete all activities as though they are critical.

Even though no float may be shown on the bar chart, astute contractors may be able to deduce that float is available. For example, Nardan Constructors can see from the bar chart that a three-week break separates completing *Erect floor forms* and starting the *Place concrete* item. If Nardan follows project procedure to keep with the early start schedule, the firm will need to find work for the concrete crew, possibly at another location, for that three-week period. If there is no other work location, Nardan must decide what to do with this crew. If the crew is made up of experienced personnel, Nardan may not be too happy about releasing them. Also, replacing them with an equally qualified crew may be difficult. The bar chart cannot solve this potential problem, but its information, brought out early enough in the project, allows the decision makers some additional time to try to resolve the difficulty.

The compressed (or summary) bar chart (Figure 9.10) displays all the activities of the Southfork Building Project on one sheet. This has an advantage over the daily bar chart, since LCC can easily scan the entire project, viewing potential

```
PROJECT MANAGEMENT                          SOUTHFORK BUILDING PROJECT                    SOUTHFORK, PENNSYLVANIA
        REPORT TYPE :COMPRESSED PERIOD BARCHART                       PRINTING SEQUENCE :Node/Activity Sequence
                                                                      SELECTION CRITERIA :ALL
        PLAN I.D.  :SBP      VERSION 19                               TIME NOW DATE      : 4/MAR/96
==============================================1996==================================================1997=================
        PERIOD COMMENCING DATE        !4   !1   !6   !3   !1   !5   !2   !7   !4   !2   !6   !3   !3   !
        MONTH                         !MAR !APR !MAY !JUN !JUL !AUG !SEP !OCT !NOV !DEC !JAN !FEB !MAR !
        PERIOD COMMENCING TIME UNIT   !2   !22  !47  !67  !87  !112 !132 !157 !177 !197 !222 !242 !262 !
=================================================================================================================
 1- 18 BE BUILD WALL FRAMES      !=====!.....!.....!.....!.....!.....!...  !    !    !    !    !    !    !
18- 28 BE ERECT FRAMEWORK        !     !=====!===..!.....!.....!.....!.....!.....!.  !    !    !    !    !
20- 21 BE BUILD MASONRY WALLS    !     !    !    !    ! CCC!CCC  !    !    !    !    !    !    !    !
21- 23 BE CHECK WALLS            !     !    !    !    ! CCCC!      !    !    !    !    !    !    !    !
-----------------------------------------------------------------------------------------------------------------
26- 28 BE INSTALL PLUMBING,MECHANICAL ! !    !    !    !    =!=====!==...!......!.  !    !    !    !    !
23- 28 BE INSTALL WALL ELECTRICAL  !   !    !    !    !    =!=====!===...!......!.  !    !    !    !    !
25- 27 BE INSTALL CEILING ELECTRICAL ! !    !    !    ! C!CCCCC!CC   !    !    !    !    !    !    !
27- 28 BE INSTALL CEILING          !   !    !    !    !    ! CCCCC!CCCCC!C   !    !    !    !    !
-----------------------------------------------------------------------------------------------------------------
28- 29 BE COMPLETE INTERIOR WOODWORK ! !    !    !    !    !    !    !    !CCCCC!CCC  !    !    !    !
29- 31 BE PAINT INTERIOR,EXTERIOR  !   !    !    !    !    !    !    !    ! CCCCC!CCCC !    !    !
30- 32 BE CLEANUP SITE             !   !    !    !    !    !    !    !    ! =====!===. !    !    !
32- 33 BE OBTAIN FINAL APPROVAL    !   !    !    !    !    !    !    !    !    !    ! CC!C  !    !
-----------------------------------------------------------------------------------------------------------------
 1- 14 KA PROCURE STRUCTURAL STEEL !=====!===...!.....!.....!..   !    !    !    !    !    !    !    !
14- 15 KA ERECT STRUCTURAL STEEL   !   !    !    ! ===!===...!.    !    !    !    !    !    !    !
 7- 15 KA PROCURE ROOF DECKING     !   !    !    ! ===!==....!.    !    !    !    !    !    !    !
15- 16 KA ERECT ROOF DECKING       !   !    !    !    ! ====!==...!.    !    !    !    !    !    !
-----------------------------------------------------------------------------------------------------------------
16- 17 KA PLACE ROOFING            !   !    !    !    ! ==..!..    !    !    !    !    !    !    !
 1-  2 NA EXCAVATE                 !CCCC !    !    !    !    !    !    !    !    !    !    !    !
 1-  6 NA PROCURE REINFORCING STEEL !=====!.....!    !    !    !    !    !    !    !    !    !    !
 1-  4 NA BUILD FLOOR FORMS        !=====!.....!    !    !    !    !    !    !    !    !    !    !
-----------------------------------------------------------------------------------------------------------------
 2-  8 NA INSTALL UNDERGROUND PIPING ! CC!CCCCCC!CCC  !    !    !    !    !    !    !    !    !    !
 6-  8 NA PLACE & TIE REINFORCING STE ! ==!====..!....  !    !    !    !    !    !    !    !    !
 4-  8 NA ERECT FLOOR FORMS        !   ! =====.!....  !    !    !    !    !    !    !    !    !
 8-  9 NA PLACE CONCRETE           !   !    ! ===!=.   !    !    !    !    !    !    !    !
-----------------------------------------------------------------------------------------------------------------
11- 12 NA CHECK FORMS              !   !    ! CCC!C     !    !    !    !    !    !    !    !
12- 13 NA CHECK FLOOR              !   !    ! !CCC  !    !    !    !    !    !    !    !    !
 9- 10 NA REMOVE FORMS             !   !    ! !==.  !    !    !    !    !    !    !    !    !
=================================================================================================================
Barchart Key:-  CCC :Critical Activities  === :Non Critical Activities  NNN :Activity with neg float  ... :Float
```

FIGURE 9.10   Project bar chart for the Southfork Building Project

trouble spots and how activities relate to each other on a time scale. The chart also adds credibility to the project by depicting graphically the grouping of the activities chronologically and according to contractor responsibility.

Milestone Reports.   Milestone reports summarize the plan and schedule and review the total program from a management-by-exception approach to quickly detect the critical items of the project. Milestones are selected events that are very important in achieving the objectives. These key events are usually start dates or completion dates of major phases of the project, the delivery date of a vital equipment item, or the date of a key management decision to ensure a successful completion date. They may or may not be critical to the schedule. When displayed on the planning diagram, the milestone data are readily available for computer input. Milestones also provide supplemental information when planning and analyzing the project.

```
PROJECT MANAGEMENT                      SOUTHFORK BUILDING PROJECT                    SOUTHFORK, PENNSYLVANIA
    REPORT TYPE :MILESTONE                                             PRINTING SEQUENCE  :Earliest Activities First
                                                                       SELECTION CRITERIA :MILESTONE
    PLAN I.D.   :SBP      VERSION  14                                  TIME NOW DATE      : 4/MAR/96
==================================================================================================================
        MILESTONE                      EARLIEST EVENT               LATEST EVENT
        DESCRIPTION                         TIME                        TIME
==================================================================================================================
   1-  1 START PROJECT                      4/MAR/96                    4/MAR/96
  13- 13 COMPLETE EXCAVATION & CONCRETE    17/JUN/96                   17/JUN/96
  15- 15 COMPLETE STRUCTURAL STEEL         15/JUL/96                   15/JUL/96
  26- 26 COMPLETE BUILDING EXTERIOR (SHELL) 29/JUL/96                  23/SEP/96
------------------------------------------------------------------------------------------------------------------
  33- 33 COMPLETE PROJECT                  10/FEB/97                   10/FEB/97
==================================================================================================================
```

FIGURE 9.11   Milestone report for the Southfork Building Project

After generating the initial milestone report (Figure 9.11), subsequent milestone reports, when produced at the same time as revisions are made to the schedule, will display quickly any departures from the original schedule. Reviewing the WBS or the planning diagram in conjunction with the milestone report will reveal activities leading to the milestone that may be delaying the project. Further analysis of their durations or scheduling dates should uncover the causes of any delays. This approach is used to quickly get to the "root of the problem."

Updating the project work activities will automatically adjust the milestones. Inputting timing changes to activities will automatically correct the milestones to reflect the new schedule.

Milestone reports are usually added to the status reports presented to the owner. Combined with the commentary, this report provides an adequate summarization of the project status.

PROJECT CONTROL (MEASURING PERFORMANCE)

As construction manager, LCC will set up the practices, including the project management procedures, for maintaining satisfactory project performance. The CM usually reports progress monthly. Contractors on the job will submit their reports to the CM, who combines and summarizes them for the progress report. The incentive to prepare this document promptly at the end of each month is that performance payments are based primarily on the progress reports.

Contractors may also prepare daily progress reports for the construction manager, depending on the critical nature of the project. This will keep LCC informed of the project work on a more current basis. The monthly report is really a composite of the daily progress reports, placing the work in a proper perspective. Contractors will include in the daily progress reports details of their work force, descriptions of work under way, and problems being encountered on the job. Figure 9.12 shows a sample progress report.

Daily Report on Progress of Building

Location _____ Date _____

Weather _____

Building _____ Job No. _____ Temp _____

Contractors	No. People	Contractors	No. People

Remarks—State in full progress of work. _____

_____

_____

_____

_____

_____

_____

_____

_____

_____

_____

_____

Delinquent contractors—State in full when notified to be on job, etc. _____

_____

_____

_____

Drawings required _____

Drawings received _____

Contractor: _____

FIGURE 9.12   Daily progress report

Equipment and Material Deliveries

Although contractors may be diligent and efficient on the construction site, actual progress may depend on the timely delivery of material and equipment. Contractors often complain about suppliers and vendors not meeting promised delivery dates. Reasons given for delays are innumerable. In any event, the construction manager needs to devise ways to control delivery dates. One way is to have the major suppliers prepare a fabrication and delivery schedule that would be compatible to the project schedules developed by the contractors. Using these schedules as a guide, the CM will prepare and maintain an up-to-date material and equipment status report. Suppliers will send periodic information on the status of the equipment. Figure 9.13 illustrates a typical equipment status report.

LCC advises suppliers and subcontractors of delivery dates, starting dates, and completion dates based on the earliest start schedule. The earliest start schedule allows for LCC to maintain complete control of the float times of the project activities. For example, if a supplier projects that a specific piece of equipment will not be shipped on the promised date, the affected subcontractor cannot start on the predetermined starting date, possibly delaying the completion date. From the early start schedule, LCC will know how much time can be tolerated before these types of deviations will affect the planned completion date of the project.

Besides the timing aspects, suppliers' contracts include specific provisions to ensure the quality of the product:

- The supplier will guarantee all work against defective material and workmanship for a specified period after completion of the work and acceptance by the owner.
- The supplier, at its own expense, will correct the defective material and workmanship.
- The supplier will submit a complete set of drawings to the construction manager for approval a specified period before the start of installation.

Project Material and Equipment Status Report

Project Title_____Trade_____Date _____

Material or Equipment	Vendor P.O. No. and Date	Date Required	Drawings To A/E Submitted	Approved	Scheduled Delivery Date	Date Checked	Percentage of Material Delivered	Remarks

FIGURE 9.13   Project material and equipment report

- The supplier will include in the original proposal the cost of any premium or overtime required to meet the completion date.
- The supplier will provide unit prices for extra work that may be authorized as not covered in the original contract.

Included among other concerns on construction projects are: (1) Changes in field conditions, (2) discrepancies in engineering drawings, and (3) modifying equipment details during installation. These situations create additional work for which the contractor will demand funds over the stated contract amount. While some additional costs are anticipated, the construction manager needs to control the extra cost disbursements. From a forecast of added funds set up at the beginning of the project, specific rules are used to control the extra costs:

- The contractor must make a written notice of a claim for increased compensation for alterations or additions before there will be any consideration by the owner.
- The contractor must show that all additions or deletions from the contracted work have written authorization from the construction manager.
- Before beginning any alterations or additional work on the original order, the contractor must receive a written order from the CM to proceed with the work. The written order, known as a "Field Authority" or "Bulletin," will contain all the information associated with changes or additional work.

Because there can be numerous extra work items, the CM needs to maintain a thorough record of all of the extra work. Figure 9.14 is a typical record. This type of record becomes useful when preparing the monthly payments and for cost forecasting.

A very important benefit in using written reports is awareness. When contractors become aware that the construction manager maintains active communication with them and the owner through these reports, they will have that extra incentive to complete their work on time and within budget. They want to maintain the reputation of being a dependable contractor, and be invited to bid on subsequent projects.

Field Authority/Bulletin

No.	Received (Date)	Returned (with Quote) (Date)	Approved (Date)

FIGURE 9.14 Format for keeping record of additional costs

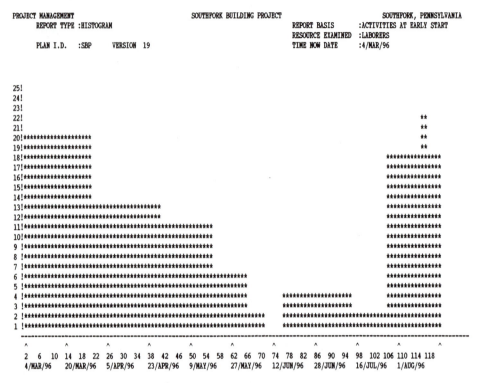

PROJECT MANAGEMENT
   REPORT TYPE :HISTOGRAM
   PLAN I.D.   :SBP   VERSION 19

SOUTHFORK BUILDING PROJECT

SOUTHFORK, PENNSYLVANIA
REPORT BASIS   :ACTIVITIES AT EARLY START
RESOURCE EXAMINED :LABORERS
TIME NOW DATE   :4/MAR/96

FIGURE 9.15   Early start schedule of laborer requirements for the Southfork Building Project

RESOURCE PLANNING

While developing a plan of action, the construction manager prepares with the contractors the resources required to complete the project. While each contractor assumes the responsibility for making adequate resources available, the CM wants to avoid conflicts among the contractors competing for the same skilled personnel. It is in the project's best interests to make an initial comparison of required and available resources for the total project.

The contractors identify their major resources on the planning diagram. The personnel most needed will be laborers, ironworkers, carpenters, equipment operators, electricians, and mechanical piping personnel. Their allocations for the individual activities are identified on the project planning diagram. The CM needs an additional diagram showing total costs, in addition to the other resources. For a number of reasons, this diagram may not be made available to the contractors. These costs include additional funds for extra work and other contingencies. Because of the confidentiality of this planning diagram, LCC will produce a planning diagram, less cost data, for contractors' use.

During the resource planning exercises and discussions with the contractors, LCC learned that the most critical supply of personnel in the area will be laborers. Since laborers will be in great demand at the outset of the project, their

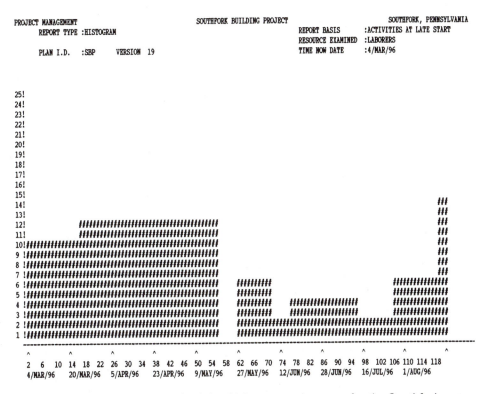

PROJECT MANAGEMENT   SOUTHFORK BUILDING PROJECT   SOUTHFORK, PENNSYLVANIA
        REPORT TYPE :HISTOGRAM   REPORT BASIS   :ACTIVITIES AT LATE START
                                                                                 RESOURCE EXAMINED :LABORERS
        PLAN I.D.   :SBP   VERSION 19   TIME NOW DATE   :4/MAR/96

FIGURE 9.16   Late start schedule of laborer requirements for the Southfork Building Project

requirements compared with supply need to be examined well in advance of the start of the project.

Figures 9.15 and 9.16 show the load charts of the laborers requirement using both the early start and late start schedules for the first half of the project. Figure 9.15 shows, using the early start schedule, that 20 laborers will be needed at the beginning and at the early stages of the project. The requirements are then diminished, so that by mid-July only two laborers are required. (During one week of that same period, the project requires no laborers.) By late July, the demand jumps to 20. This inconsistent requirement pattern for laborers is not too desirable. LCC needs to examine the laborer load chart using the late start schedule.

The late start schedule, shown in Figure 9.16, displays improved loading for laborers. A constant requirement of 11–12 laborers lasts for the first six weeks, and while the next two months see peaks and valleys in laborer demand, the contractors may prefer using this schedule. On the other hand, the CM wants to use the early start schedule. (LCC loses some control over the project schedule by using the latest start schedule since all activities have no float and any drift of any one activity can delay the project completion.) Each contractor will review and then report on its ability to handle the inconsistencies before LCC makes a final decision.

CONTROLLING PROJECT COSTS

LCC must conduct business with a full knowledge of the status of the project's costs. The need to negotiate temporary loans from banks or establish credit to pay suppliers is dependent on controlling the funds allocated for this project.

After substantiating the costs for each project activity, LCC prepares a cost schedule showing how these costs are distributed over the project. Whether to report on a monthly, weekly, or quarterly basis depends on the critical nature of funds. Cost distribution allows LCC to plan its cash flow over the length of the project. LCC can also use the same cost distribution data as a basis for measuring cost performance.

LCC starts the procedure for controlling costs concurrent with the planning process. The total costs of completing each project activity are placed on the project planning diagram (previously noted in Figure 9.5).

As the cost histogram is a daily report, LCC will need to express the input cost data for each activity in dollars per day. The cost histogram, also known as a cost distribution graph, shows total costs expended on a daily basis. Its use is effective for reviewing the cost schedule.

LCC prepares the cost distribution graph shown in Figure 9.17. The graph reveals an outlay of $60,000 daily for the first three weeks of the project, reducing to the mid-$40,000 range for the next three weeks. In six weeks, one-third of the total project cost of $4,450,000 is spent, and only 10% of the project is completed. Performance payments will be less than the contractors' cash outlay, meaning that the contractors will need to seek additional funds at the early stages of the project to pay the labor, material, and equipment costs. They prefer that the schedule be set up so that performance payments will finance their costs as the project progresses.

If these seemingly large expenditures at the early stages of the project cause serious potential cash flow problems to the contractors, LCC would evaluate the possibility of adjusting project jobs that have float availability. By comparing the financial impact of both early start and late start schedules, LCC prepared accumulated cost reports of both schedules for the March 4 through August 8 period. The computer-generated reports of these schedules are shown in Figures 9.18 and 9.19.

The cost curve reflecting the late start schedule clearly shows fewer expenditures over the first five months of the project—$3.6 million versus $3.0 million. Tradeoffs of a tighter schedule (latest start schedule) or higher initial costs (early start schedule) are to be evaluated. The time to consider a schedule strategy change is in the planning stage. If the latest start schedule is adopted, the contractors need to be advised that the schedule is tight and no changes can be tolerated. Few deviations will result in a serviceable control of timing and costs.

The Indicated Cost Outcome Report

For effective cost control, LCC uses an indicated cost outcome report to help contain project spending within approved (authorized) amounts. LCC uses the

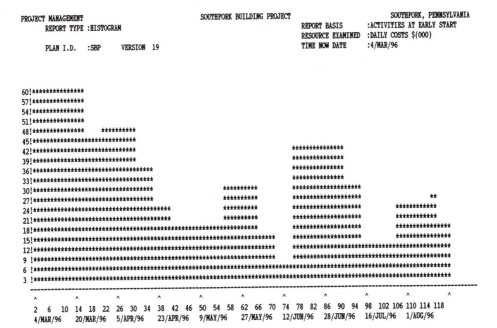

FIGURE 9.17 Early start schedule cost distribution graph for the period of March 4 through August 8

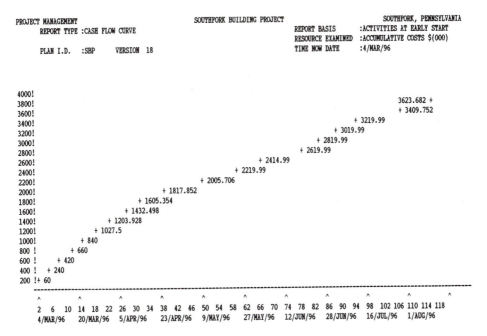

FIGURE 9.18 Accumulative cost curves of early start schedules for the Southfork Building Project, April 4 through August 15

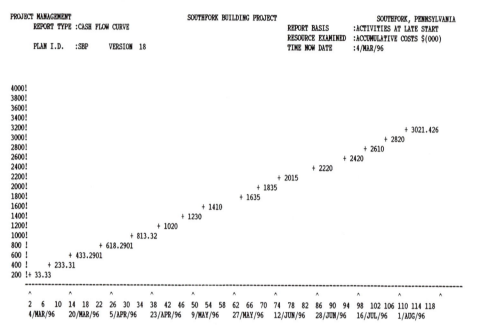

FIGURE 9.19   Accumulative cost curves of late start schedules for the Southfork Building Project, April 4 through August 15

report to (1) review and evaluate the spending status of projects, (2) determine if project commitments are in line with authorized amounts, and (3) determine if (and when) additional authorizations may be required.

The indicated cost outcome format, simple to develop and apply, is usually prepared monthly. (Chapter 5 discusses the indicated cost report in detail.) Prepared with a spreadsheet package, the main component of the report is the amount of work that is yet to be completed, and at what cost.

CONCLUSION

This chapter emphasizes the need for thorough planning as the first of several steps in achieving successful results, using as an example the Southfork Building Project. For this type of project, assigning a construction manager (CM) to head the program is an excellent move. The CM will immediately take steps to execute the project by: (1) developing procedures to administer the project, (2) forming the office and field organization, (3) planning the project, (4) preparing budgets, and (5) preparing the master plan and schedule.

The CM and the selected contractors will prepare their respective plans, and in joint discussions and meetings will complete the master project plan and schedule.

In developing the plan and schedule they use these guidelines:

- Plan the project to balance cost and time, and avoid excesses in labor, material, equipment, and any other required resources.
- Plan and schedule on the basis of past experience on similar projects. Recognize that you will need to replan and reschedule on a continual basis because of unforeseen conditions.
- Understand the importance of planning. The planning diagram, a graphical portrayal of the plans for carrying out the work, is proven as an effective planning expedient.
- Develop a workable plan of the project work items that make up the project. This plan includes a WBS with clear project activity descriptions, activity interrelationships, and the resources to be assigned for each activity.
- Schedule the activities within the agreed-upon time span.
- The dominant guideline: One should first plan and then work the plan.

A completed plan identifies accountability and recognizes the work each contractor must complete. It also permits the CM to successfully conduct and control the project. In controlling the project, the CM measures performance using reliable techniques, such as the project timing bar charts and the indicated cost method. (These and other control procedures are explained in Chapters 4 and 5.)

The Southfork Building Project, like all construction projects, faces inevitable changes. Anticipating the effect of changes on the project and how to convincingly deal with them is inherent in the project management techniques that are used. With the aid of the computer, changes causing problems that are found early in the project can usually be resolved before a crisis occurs.

Exercises

1. The present construction plan allows for completion of the project on February 10, 1997. The City of Southfork has asked the contractor, Lawrence Construction Co., to reduce the completion time by three weeks.
   a. What changes in the plan could be made to reduce the duration? (Check the critical project items: Can any of them be divided into two parts and done concurrently with other items? One example is to do three weeks of the *Painting* activity concurrent with the *Complete woodwork* item and reduce the duration of the *Painting* activity to three weeks. Use the screen report of the software program to note changes and check results. If they are favorable, generate a print report. Are there any other possibilities?)
   b. Assuming that the plan is to remain intact, what timing changes might be considered to reduce the project completion date by three weeks? (Suggestion: Review the computer-generated critical activities schedule for the critical activities that occur early in the schedule. Could several of

them on the same critical path be reduced by one week each by overtime or adding personnel?)

c. Prepare a report supporting your decisions and include the new schedule.

2. On March 18, Nardan informs LCC that several of its activities will be finishing later than originally planned:

Activity	Planned Finish	Revised Finish
Build floor forms	March 29, 1996	April 20, 1996
Install underground piping and electrical	May 17, 1996	May 24, 1996

Durations have changed: *Build floor forms* is extended from four weeks to seven weeks; *Install underground piping & electrical* from eight weeks to nine weeks. LCC accepted as valid Nardan's reasons for the delays. To keep the schedule intact, LCC must reassess the project plan and schedule:

a. Input new durations reflecting the revised finish dates. What is the new project duration? Has it changed from the schedule, and if so, by how much?

b. If the new project duration is longer than listed on the original schedule, which activities will be considered first for reducing their timing?

c. From the above list of critical activities, which ones would you choose to revise to maintain the original schedule?

3. When planning the personnel required for the Southfork Building Project, LCC believes that, in addition to the laborer demand, supplying carpenters and electricians for the project may also pose a problem during the May 4 through August 8 period.

a. Using the early start schedule, prepare a report showing the carpenters' and electricians' workloads for the May 4 through August 8 period.

b. If six carpenters are available for this project, will they be adequate to handle the demand?

c. If there is an overdemand, can it be handled by allotting overtime to the carpenter crews? Support your answer by marking the available supply on the carpenter load chart.

4. LCC had already reviewed the cash flow disbursements for the period of March 4 through August 8. Now the CM wants to review the cash flow from August 8 to the end of the project. Produce this computer printout for both the early start and late start schedules.

a. What significant differences do you observe?

b. Are there any good reasons to change the schedules from early start to late start on the strength of a better cash flow?

A Facility Project: Measuring Project Performance

The board of directors of Gilmore Industries approved $3,502,000 for a program to expand and improve the paint finishing line of the company's fabricating plant. The expansion will include a building to house the paint spray process and to store added inventory. To meet the customer demands of their product, Gilmore Industries must complete this project by September 24, 1996 so the company can start production by that date. Gilmore has plans to start the project on November 8, 1995.

Gilmore Industries has retained M&A, an industrial engineering and contracting firm, to prepare a plan of action that will support the completion date and formulate methods to control the project performance. M&A's presentations convinced Gilmore that it had resources to handle the project and achieve the objectives. M&A was awarded the contract to build the facility, including building design and construction as well as equipment procurement and installation.

Important features in M&A's presentation included the following:

- M&A will begin procuring the long-lead equipment items (those requiring longer fabrication and delivery times) when their designs are completed. Placing their design schedule on a higher priority will give the project a better chance of meeting its schedule.
- "Fast track" methods will be used for building design and construction. The building will be divided into

two phases: Phase I consists of the building "shell"—site, foundations, structural steel, floor, roofing, and wall siding—while phase II includes interior, mechanical and electrical utilities, painting, and approval. Bid documents will be prepared in two separate packages and contracts awarded separately by phase.

- M&A will use "fast track" methods for production equipment design and installation. Phase I includes the major equipment items that will have little or no fabrication and delivery problems. Phase II will follow with equipment that requires longer delivery time and more complex installation.
- A representative will coordinate the work of all of the building construction contractors and those installing the equipment.

MEASURING PROJECT PERFORMANCE

Gilmore Industries set the main objectives that will be the guide for developing the WBS: Build the facility for $3,502,000; start by November 8, 1995 and complete by September 24, 1996.

The three major divisions required for this facility are: (1) building, (2) equipment, and (3) training personnel to operate and maintain the building and equipment. Each of these three major divisions will require key work activities to complete:

- **Building** consists of the facility design and construction.
- **Equipment** requires the design, installation, tryout, and launch of the equipment.
- **Training** consists of preparing training manuals and selecting and training qualified personnel to operate the new facility.

M&A, upon the notice to proceed on November 8, 1995, established objectives and some significant milestones that included the following:

Objectives and Milestones	Date
Start project	November 8, 1995
Start equipment design	November 8, 1995
Complete building design	January 30, 1996
Complete phase I construction	May 21, 1996
Complete equipment installation	September 10, 1996
Complete project	September 10, 1996

Preparing the Work Breakdown Structure (WBS)

M&A assigned a supervisor to head the three major aspects of this project—building, equipment, and training.

The project team went through several breakdowns of activities before completing the WBS. The team finally completed the master WBS, which organized the pertinent work activities in chart format, shown in Figure 10.1.

Gilmore Equipment Installation Project
Work Breakdown Structure

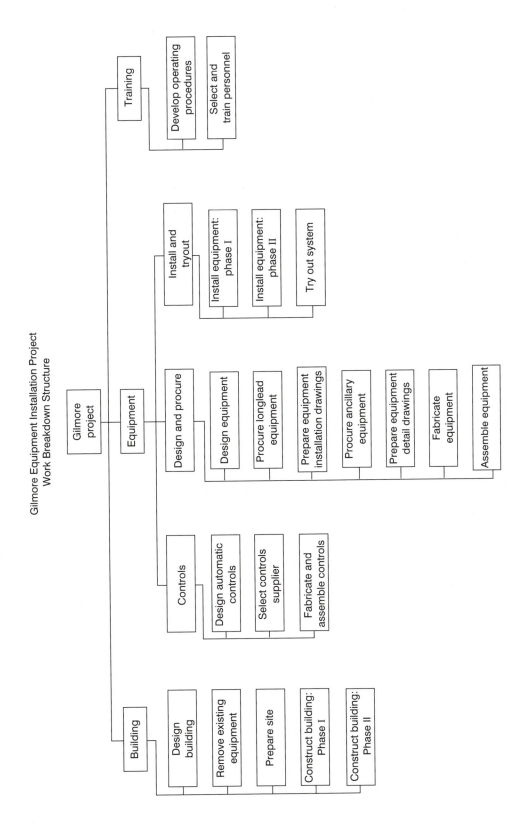

FIGURE 10.1 (Master) work breakdown structure for the Gilmore Equipment Project

DESIGN AND CONSTRUCT BUILDING

M&A decides to relieve its work load by subcontracting the building design and construction to a design/build firm. (Because of M&A's expertise, it has chosen to concentrate on the equipment and training for this project.) As shown graphically in Figure 10.2, four major sections characterize the extent of the work needed to complete the building division:

- Design building: User requirements, site requirements, building design, site design, building specifications, bid processing, and contract award.
- Prepare site: Earthwork, grading, underground installations (electrical, water, and sewer lines), and roadways.
- Construct building—phase I: Foundations, structural steel framing, roofing and siding, and floor construction.
- Construct building—phase II: Interior work, mechanical and electrical systems, lighting, painting, and approvals (including cleanup).

M&A asked Lawrence Construction Company (LCC) to prepare a proposal to design, prepare the site, and construct the building in accordance with specific instructions. These instructions were documented in the form of "packages." One of these is the user requirements, which are basic data that Gilmore Industries prepared to provide an understanding of its needs. Typical user requirements include a plant layout showing location of equipment and other major building items; total area of the building; bay sizes (area between building columns); ceiling heights (truss heights); electrical power, water and fuel schematics, and special foundations that may be needed.

The user requirements are the first data to be provided. LCC and Gilmore will continue to furnish data as design development continues. A series of review and approval meetings continues as designs are completed.

(The A&E and construction work discussed in Chapters 8 & 9 is typical of the work LCC will do to complete the design and construction for the Gilmore Equipment project.)

BUILDING

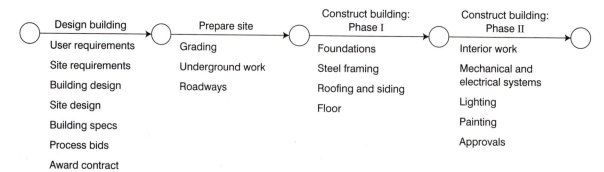

FIGURE 10.2   Subdiagram (building) for the Gilmore Equipment Project

DESIGN, INSTALL, AND TRY OUT EQUIPMENT

The main objective in the equipment section is furnishing quality equipment to produce high-quality parts for Gilmore's customers. Of equal significance is timely completion of all of the work associated with the equipment process.

The work associated with this section includes the following: (1) preparing the total equipment plan, (2) preparing equipment designs and a specification package, (3) soliciting and reviewing supplier proposals, (4) preparing equipment installation designs and specifications, (5) fabricating, assembling, and delivering equipment, (6) soliciting and reviewing installation contractors' bids, and (7) installing and trying out equipment. During this period there are a great number of meetings with engineering and operating personnel, who, although they may not be directly involved with the project, will nevertheless be helpful in equipment selection through their knowledge of the operation. Informal contacts with suppliers will also be helpful. While time-consuming, meetings with all knowledgeable persons are worthwhile. These experiences provide the impetus to start procuring equipment with some confidence. It is important to make allowances in the timing and workload analysis to perform this necessary function.

The Equipment Plan

As a guide to direct the work of the equipment section, M&A prepared a partial planning diagram (sometimes referred to as a subdiagram), that will become part of the project planning diagram. This subdiagram (Figure 10.3) is a result of meetings and discussions with personnel who may be participating in the project, including potential suppliers and contractors.

The equipment section is the largest group. It will include major sections: Design and procure, Controls, Install, and Try out. Each of these sections is subdivided into a number of individual work items.

- Design and procure:
  a. Design equipment (equipment layout, process description, design requirements); prepare equipment detail drawings (process proposals, select suppliers); fabricate equipment; assemble equipment.
  b. Prepare equipment installation drawings (solicit proposals, select contractor); remove existing equipment (clear existing area).
  c. Procure special equipment—long-lead items and ancillary equipment.
- Install and try out:
  a. Install equipment—phase I (install major equipment, install ancillary equipment, reinstall acceptable existing equipment).
  b. Install equipment—phase II (install long-lead items, install controls, make electrical and mechanical connections).
  c. Try out system (try out individual equipment, try out total system, launch product).
- Controls:
  a. Design controls.
  b. Select controls suppliers.
  c. Fabricate and assemble controls.

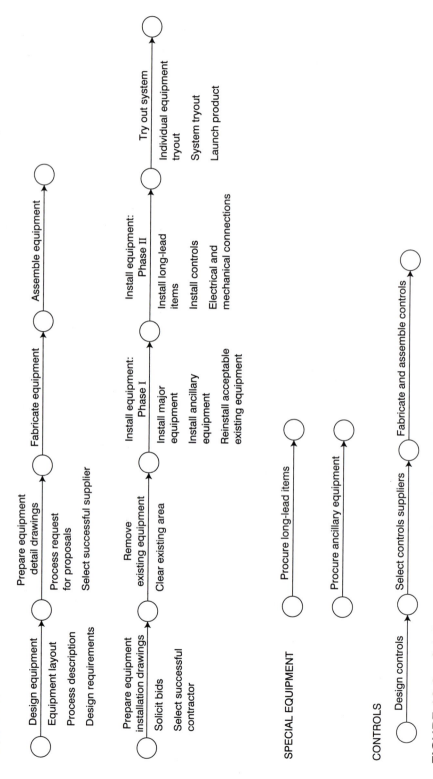

FIGURE 10.3 Subdiagram (equipment) for the Gilmore Equipment Project

Design and Procure Equipment.   The initial step in the equipment process is for M&A to prepare a detailed plant layout of the new equipment (and the existing equipment that will be relocated). Included in the plant layout will be the positioning of storage areas, maintenance shop areas, plant offices, aisles— all the operations essential to production. The plant equipment layout serves as a valuable tool and guide in preparing equipment designs and specifications. The layout also determines the overall building dimensions needed to adequately enclose everything. Other important building design aspects include locations of electrical, water, and fuel services; building bay and aisle sizes; and ramps and scale pits for shipping and receiving.

Supplementing the plant layout is a set of process requirements that explain the equipment's functions. The process requirements follow each operational step that culminates in the creation of the product.

As commonly practiced in many industries, M&A will submit the plant layout and process requirements to specialized engineering and design firms to provide the specific engineering and design details needed for fabrication and assembly. Another source for designs and design details is the successful equipment supplier. Frequent meetings allow review and approval of the equipment designs and details prior to starting fabrication.

Processing equipment proposals can be time-consuming, so the schedule needs to allow adequate time to solicit and review proposals. Some complex equipment suppliers may need anywhere from three to eight weeks, with an average of four weeks, to complete their proposals. M&A may need almost an equal amount of time to review the legal aspects of the proposals and prepare comparison tabulations of the bid proposals. Investigating the financial credibility and technical competency of the proposed suppliers may be required on certain occasions. (If a large number of suppliers request to submit proposals, prior evaluation to select those most qualified may be advisable.)

Suppliers may include in their proposals unsolicited suggestions for improvements, modifications, or even a request for clarifications. If these have some validity, it is in the project's best interests for M&A to consider. Bidding time will be extended for meetings for all prospective suppliers to clarify parts of the bid documents. Separate meetings will be held with bidders offering suggestions and modifications that may have an impact on their individual bids. It is also protection to the bidder who offered the ideas.

M&A recognizes that time is a factor, so the firm elects to have the successful bidder prepare the equipment design and details. With this provision in the bid documents, all bidders will have access to the equipment layout, process description, and equipment design requirements to prepare their proposal.

The time allowed for the bidders to submit their proposals should be enough to provide adequate reviews and analyses. M&A cannot minimize the importance of the selection process and it needs to be shown in any plan of action. It is commonplace among equipment suppliers to have unique features of their product. Inevitably they will supplement the specifications with suggested modifications and improvements that they consider worthwhile.

Fabricate, Assemble, and Deliver the Equipment

Among the first items of discussion with the successful equipment suppliers are delivery schedules and how they are designed to complete the project. While the bid documents will state delivery times, they are reviewed again with special emphasis on the most critical equipment deliveries. It is most important to discuss with equipment suppliers the meaning of early start schedules. Unless the importance of meeting commitments is emphasized, equipment suppliers tend to be permissive in delivering their material and equipment.

Equipment suppliers will emphasize their ability to combine quality and cost when selling equipment. Highly qualified engineers skilled in specific disciplines with a combined talent can produce a competitive product, but they may lack project management skills, especially in planning and scheduling functions. M&A has included in its initial discussions preparation of comprehensive schedules for fabricating, assembling, and delivering the equipment. (If there is doubt about the supplier's abilities, then M&A will offer succinct instructions on their preparation.)

Equipment Control Systems.

When control systems are complex, selecting a controls supplier by price alone may not be the correct approach. It is important to get separate control systems proposals for relatively complex equipment, as they will normally require equally complex controls.

Preparing separate controls specifications can become quite intricate. A compromise arrangement is to have equipment bidders specify a special control systems supplier in their bids; that supplier will then provide the proposal for the electrical and other systems of the equipment package.

What makes the control systems proposals difficult to evaluate is that different programs may be needed to operate the suppliers' equipment. Evaluating and comparing the proposals may thus be a lengthy process. The accepted proposal may necessitate altering the equipment specifications and require additional bidding. In another scenario, only one supplier in prebid discussions may be able to meet the specifications. In this type of situation, all the equipment bidders would be instructed to include this controls supplier and proposal in their bids.

The bottom line when planning and scheduling controls is to treat them as a separate part of the plan. Knowing that they are complex and poorly understood, allow adequate time for design, selection of suppliers, and fabrication.

Equipment Installation Drawings

Designs for equipment installation use the plant layout drawings to designate the equipment's exact location. Included in equipment installation drawings are layouts of the equipment foundations, routings of the mechanical and electrical distribution of the utility piping, and special details associated with installing the equipment.

The details in installation drawings are very helpful in preventing physical interferences within the plant. Good installation designs will also provide proper pipe sizing and adequate electrical conduit and wiring arrangements to produce a completed installation. Design engineers will prepare the proper sizing and so forth, and CADD personnel will show all of these details in the installation drawings. Relying on the skilled personnel to do the work in the field for designs and details (a frequent practice) is risky. Allowing field design work carries the danger of no records; when references are needed at later stages, there may be none.

M&A will furnish a coordinator to oversee exchange of information between equipment designers and equipment installation designers to organize the installations. Inevitable equipment design changes may affect foundation dimensions and utility services being designed by the equipment installation designer. Providing these types of design changes to the equipment installation designer promptly will avoid costly changes and potential delays during building construction and equipment installation.

On occasion the equipment installation designer may note discrepancies in the equipment design that should be relayed to the equipment designer. Almost all communications between the equipment design and equipment installation design functions should be through the coordinator. Most engineering and design offices require documenting confirmation of changes and related matters. Depending upon their impact on the total design, they could become part of the agenda at regularly held progress meetings.

Installation drawings and specifications are prepared in prescribed sections. For example, the general contractor should be able to separate the mechanical and electrical sections to get specialized subcontractors to bid on their respective sections. Preparing thorough installation drawings and specifications helps avoid uncertainty. Bidders who can understand the drawings and specifications will prepare reasonable proposals, permitting consistent evaluation and acceptable bid pricing.

Procedures for soliciting equipment installation design bids and their subsequent reviews are similar to the equipment supplier bid process. Contractors that will bid on this type of work are generally those who do specialized work in building construction and are familiar with installations of building service equipment. Sometimes the production equipment installations are more complex and sophisticated, and therefore require more specialized contractors. Equipment suppliers are a good source of qualified bidders, since they have probably worked with them on other jobs and are familiar with their performance installing their equipment.

Supervising technically oriented installations using highly skilled tradespeople requires special attention to ensure quality. In the Gilmore project, M&A recruited a qualified person from the equipment supplier to be the field supervisor for the installation work. He will act as an adviser to the contractor. With this arrangement, the contractor is still responsible for the work and all of its activities, including assigning and directing the skilled trades personnel, ordering the material, and handling all administrative duties.

The specifications will define who is to pay for the services of a field supervisor. Specifications need not be as detailed when the owner or owner's representative pays for these services as when the contractor does. In the latter, specifications will detail the scope of the assignment, the responsibilities of the equipment supplier's representative, and the time period for the services.

Qualified representatives serve two purposes: (1) the owner will be satisfied that the work is completed in accordance with the supplier's designs, and (2) from the specifications, all bidders will be aware of the scope of services the supplier's representative will provide. Consequently, the cost of this item should be consistent among the bidders.

Equipment Installation and Tryout

Equipment installation starts when phase I building is essentially completed, and adequate equipment has been delivered to permit installation to start. The installation contractor may already have mobilized on the site to receive delivery of the equipment prior to the start of the installation. The installation coordinator will have arranged for completed parts of the building to be reserved for equipment storage when they are unloaded. This part of the building will be used exclusively for equipment storage until installation begins.

The installation contractor should be completing or have completed by this time a detailed plan and schedule of the equipment installation work. Reviewed by M&A, its approval will be based on meeting the project objectives. M&A will formally monitor this schedule at least monthly, assessing the critical items on their impact of the completion date and, depending on their seriousness, resolving any problems with the contractor before any crises occur.

Equipment tryout can start upon completion of the building utilities in *Construct building phase II* and the equipment installation included under *Install equipment phase II*. Contractors will try out individual equipment before total system tryout to minimize and isolate the problems that may occur when the entire system is started. Initial tryouts of individual equipment will net an earlier, more successful launching of the total process system.

At equipment tryout, electricians will do such tasks as starting motors for proper rotation and checking balancing and vibrations. Electrical technicians will check individual electronic controls, computer programming, and sensors for adequate performance. Mechanical personnel will check valves for control of the water, natural gas, and compressed air supply to the equipment.

It is advisable to have plant personnel perform the tryout tasks. If, by contract, contractor personnel are responsible for ensuring that equipment operates satisfactorily in this tryout period, Gilmore should still authorize its own skilled personnel and selected supervision to observe. Also present should be persons being trained to operate the equipment when production begins. For production, the tryout period is the most critical time, as potential equipment malfunctions will appear that may become chronic quality problems causing costly delays. Permanent plant personnel need to be present to learn the symptoms and possible solutions.

Total System Tryout and Launch

Participants at the system tryout need to keep a record of their experiences—especially causes of problems and solutions—with equipment performance. These will be an invaluable reference during the system tryout and launching. (Launch starts when the first part goes through the system and is completed when the full complement of parts flows through with no quality problems. Production officially starts at that time.) In addition to the log, plant personnel will prepare "punch lists" (malfunctioning items that need to be corrected by the equipment installation contractor). Equipment corrections need to be completed during equipment tryout, system tryout, and launch. Other matters will cause enough problems when production begins.

In the past, when equipment facility schedules were running late, U.S. companies had a tendency to minimize the efforts of the system tryout and launch phases. Getting into production as quickly as possible compromised the need to achieve a quality product, however. This situation was compounded when quality control persons, who reported to production supervision, approved questionable parts under pressure. They did not want to be accused of not being team players or being obstacles to meeting production objectives. The consumer in the marketplace accepted substandard products of poor quality. Then Japanese manufacturers entered the market with their quality products, especially noticeable in the automotive market. One of the underlying reasons behind their high quality was their emphasis on the tryout and launch periods.

For producing quality products, U.S. manufacturers now submit rigid standards to suppliers of production and other equipment. Quality standards in manufacturing have become rigid as well, with tryout and launch controlled by specific procedures to achieve quality equipment performance. As another safeguard, quality functions and personnel are now autonomous, reporting directly to top management.

Installation and tryout of the equipment need to be 100-percent complete before the production begins. Tight specifications are enforced. Specific equipment components that are not functioning properly must be replaced before production can begin.

Measuring Tryout and Launch Performance

The major steps of tryout and launch consist of planned sequences of equipment testing, balancing the systems, and their final acceptance. This phase will be completed in various stages over a time, the length of which depends on the complexity and scope of the project. (On a Ford-Japan joint venture parts plant in Tennessee, this phase was scheduled for six months—one-third of the total scheduled duration.)

A team leader is appointed to manage the owner's responsibilities associated with the tryout and launch phase. Because this phase may be interpreted as an initial stage of production, this person is a logical choice to be placed in charge of the facility when completed. The team leader should prepare a plan and schedule for coordinating the people and equipment used for the start-up. A plan and schedule will discourage imposed shortcuts to make up for delays experienced in the prior phases of design, fabrication, and/or installation. This document becomes all the more important in helping prevent serious problems when operations begin.

All persons associated with the project, from plant management to maintenance and operating personnel, must be informed of progress. Plant management should be informed weekly of the present status, the major problems, causes of these major problems and details on how they are to be resolved. Keeping all other affected personnel informed not only of the launch but other project matters as well is beneficial. Uncertainties about how the new facility is progressing or how it will operate may affect their livelihood (e.g., highly automated equipment could mean fewer jobs). Not handling communication properly may affect operation of the system.

As an example of poor communication, years ago, when installing a utilities monitoring and control system (long before the term "energy management system" was used to describe such installations), plant management did not inform the plant services personnel that the new automated control system would cause a relatively small reduction in personnel. As the system was being installed, rumors raged exaggerating the real effect of the control system when it became operable. Subtle, deliberate slowdowns occurred throughout the plant and remained months after the system was in operation. Anticipated savings from this facility were never achieved. Ironically, all the displaced persons were to be reassigned within the plant; plant management had not considered it important enough to inform anyone that they would not lose their livelihood.

Status meetings on launching objectives need to be held daily. Separate meetings with agenda on specific subjects with only the affected personnel present are preferable to a group meeting with too many people who have diverse interests.

Activity	Participants
• Discuss status of jobs and work to be performed	Maintenance and operating personnel
• Discuss status of total project and specific jobs; set plans for the daily jobs	Maintenance and operating supervisors, technical staff (task force), contractors
• Lay out work plan, identify personnel needs for next three days, resolve safety concerns	Technical staff, plant supervision, contractors

TRAINING

Figure 10.4 is the planning subdiagram for the major activities, *Developing operating procedures* and *Selecting and training personnel*. It also includes the supporting activities needed to complete the training for operating, maintenance, and quality personnel to run the new facility.

FIGURE 10.4
Subdiagram (training) for
the Gilmore Equipment
Project

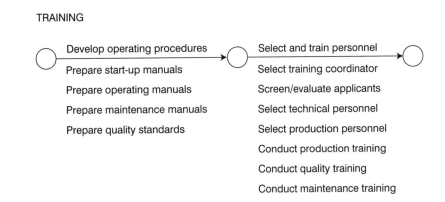

TRAINING

Develop operating procedures

Prepare start-up manuals

Prepare operating manuals

Prepare maintenance manuals

Prepare quality standards

Select and train personnel

Select training coordinator

Screen/evaluate applicants

Select technical personnel

Select production personnel

Conduct production training

Conduct quality training

Conduct maintenance training

Developing Operating Procedures

Preparing the manuals begins at the ending stages of *Design equipment* of the major *Equipment* division, and some volumes cannot be completed until after the *Tryout and launch* phase. Manuals are prepared to support operations through launch and production. Designated persons from the group preparing these manuals will select and train personnel who are qualified to operate and maintain the facility.

Four major categories of manuals are prepared: start-up, operating, maintenance, and quality standards. Documenting these divisions of operations is necessary, not only for training purposes, but for continual reference when production operations are under way. The manuals will be organized to allow separate volumes for each major subsystem and separate sections within each volume for the equipment used in operating the system. Some general reminders for their preparation include:

- Use three-ring binders that allow insertion of added material or removal of outdated data or data not usable because of changes.
- As this material will be used over a number of years and by many people, use good-quality binders and paper. (Consider using plastic protection sheets for the most frequently used and the most important information.)
- Technical specifications are complex. To make the manuals readable and understandable, use simple steps in preparing the manual: (1) what to do, (2) when to do it, (3) how to do it, and (4) why to do it.

The first materials to be prepared are start-up manuals for the training group. The group will need them to train the personnel who will be operating and maintaining the facility. (Some of the trainers may already have been assigned to work on the equipment and will double as instructors.) Start-up manuals are communication links with the persons who have designed and installed the equipment and control systems. They also help familiarize operating and maintenance personnel with the new facility.

Persons will use start-up procedures manuals beginning with the start of the tryout period and continuing throughout production. During production, start-ups will follow periodic shutdowns for maintenance purposes. Start-up manuals will include descriptions of the operating sequences of the equipment, descriptions of the equipment's functions, and balancing and testing of all of the components. The start-up manual documents are essentially an organized method for starting up the complete process.

Start-up manuals begin with the complete start-up plan of action in outline form. Using the same theme as set up for a project, objectives, and an organizational arrangement like the WBS, the start-up becomes a thoroughly thought-through effort. Start-up manuals supplement equipment design drawings and specifications and complement operating and maintenance manuals. They are especially important to prepare for automated (computer-controlled) systems.

Operating and maintenance manuals include a separate section (chapter) for each system. Each chapter describing the system includes the following subsections:

- Introduction
- Table of contents
- Description of the system
  - a. Engineering data
  - b. Flow diagrams
- Start-up procedures
- Operational procedures
  - a. Operation sequences
  - b. Emergency procedures
- Maintenance
  - a. Preventive maintenance
  - b. Corrective maintenance
  - c. Troubleshooting procedures
  - d. Special maintenance requirements
  - e. Manufacturers' literature
  - f. Shop drawings
  - g. Spare parts list
- Record filing system

Complete sets of final "as built" drawings for all building construction, fixed equipment, mechanical and electrical supply equipment, and mechanical and electrical distribution systems become part of the operating and maintenance manuals. These manuals have two parts: instruction and technical reference.

The instruction section covers specific work efforts: what to do, when to do it, how to do it, and why to do it. In addition to the as-built drawings and mechanical and electrical schematics, the technical reference includes equipment and systems descriptions, design criteria, operation and maintenance sequences, major maintenance overhaul details, parts lists, and equipment repair history.

The manuals will emphasize the importance of record keeping, especially for the equipment repair history. This type of record keeping is important to maintaining high equipment uptime. Knowledge of past inspection records (especially lubrication), descriptions (dates, times, costs) of prior maintenance, and records of repairs and visits by manufacturers' representatives concerning equipment problems are essential. Information should be recorded in a well-organized, readily usable format. Verbal transmittal of operating and maintenance events is an ineffective way of communicating with management and other plant personnel.

Manuals prepared for quality assurance and quality control specifications set the basic standards for attaining quality performance in producing the product. Not only the training personnel but all of the facility personnel will study these manuals, as they will be often told: "Quality is a matter of our company staying in business."

The quality training continues through the lifetime of production operations, and one of the expedients used is the formation of quality circles composed of all of the employees. Managers to the lowest grade employees attend these quality circles. They are "team members" becoming involved in all sorts of problems and concerns related to the operations. Through the quality circles the employees become absorbed in the decision-making process, making countless recommendations for producing a better product. Continuous improvements motivated by the quality circle meetings require the quality manuals to be constantly updated.

Selecting and Training Personnel

The plan needs to provide for timely selection and training of personnel. Some trained electrical and maintenance personnel should be present starting with the late stages of equipment installation. Additional skilled persons should be present when equipment tryout begins. A fair complement of production, maintenance, and quality personnel need to be trained prior to the systems tryout and launch period.

If there appears to be a shortage of qualified persons to start up the facility, the firm may consider recruiting qualified personnel from the equipment installation contractor's staff of electricians, mechanical, and other personnel. Since these offers can turn into permanent positions, they are often attractive to selected installation personnel. This is especially evident near the end of this project.

PLANNING AND SCHEDULING THE PROJECT

From the previously prepared building, equipment, and training subdiagrams, M&A identifies the interrelationships among these major sections in the planning diagram shown in Figure 10.5. The diagram also indicates durations of all the activities. (For computer purposes, the nodes are given identifying numbers.)

Constructing a plan of the project is analogous to preparing a layout of the facility. This plan gives M&A an excellent picture of the total project. This initial effort includes mainly major activities. At a future date, M&A will expand this diagram to include more detail. M&A will use the detailed plan, which will include all of the supporting activities shown on the subdiagrams, for directing the work of the contractors assigned to this project.

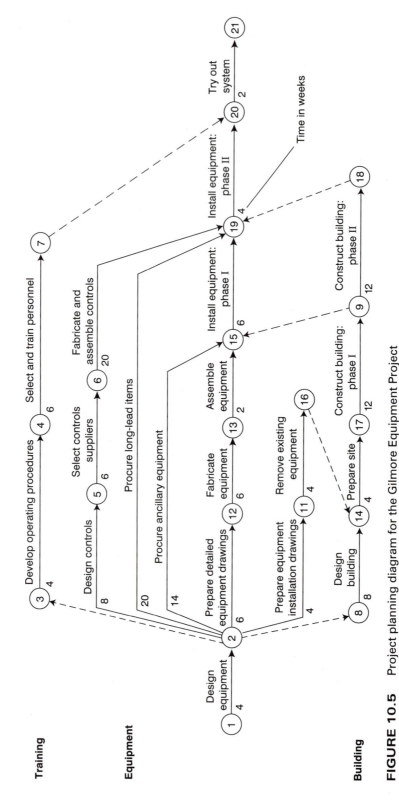

238

FIGURE 10.5 Project planning diagram for the Gilmore Equipment Project

The planning diagram developed so far can be used for preliminary analysis. M&A can do some preliminary checking of the validity of individual activity durations against the project timing objective (September 10, 1996). By using the manual calculating method, M&A can establish the earliest start times for each of the activities. To determine the earliest start time of the last node (21), which is defined as the duration of the project, the calculations made on the plan (Figure 10.6) show it to be 46 weeks after the start of the project (November 8, 1995). The plan agrees with the time objective set for the completion of the project, and it can now be used to provide other information.

M&A uses the plan to check the start of significant project activities—one of the important events or milestones of the project. For the earliest date to schedule the start of equipment installation, M&A checks the earliest start time of *Install equipment phase I* (Node 15), which is 28 weeks (April 10, 1996). With this information, M&A can advise the appropriate phase I equipment suppliers that their equipment must be received at the facility by April 10, 1996. Any dates derived from this method should be confirmed as soon as feasible by a computer scheduling program.

With suitable estimates, M&A can begin to identify and earmark skilled personnel resources and costs on the planning diagram. Figure 10.7 shows the final diagram (or master diagram). In effect, completing this diagram completes the planning phase, and scheduling the project can now begin.

Applying Computerized Scheduling

All the data needed to generate a computerized schedule is on the master planning diagram. Using a worksheet (Figure 10.8) speeds up and simplifies transfer of data to the computer input screen. The worksheet format replicates the input screen to a certain extent for node numbers, activity descriptions, durations, and resource data (excluding costs).

M&A desires reports for use in the field office. These reports include: (1) a timing schedule, (2) bar chart schedule, (3) skilled labor requirements when using the early start schedule, and (4) skilled labor requirements when using the late start schedule. The firm will also request separate printouts for the building, equipment, and training schedules. When needed, they will also generate milestones and start and/or finish dates of major events. M&A will use the milestone report to summarize project performance.

Figure 10.9(a) shows the computer-generated report listing the project schedule. M&A keeps this report for its own use as it shows the float availability. This report is not given to any of the suppliers and contractors for fear that they may abuse the float feature, using the late start schedules as a basis for their own schedules. M&A will edit this report for use by outside personnel, showing only the durations, early start, and early finish times.

M&A will distribute the bar chart shown in Figure 10.9(b) to management personnel, since it shows float availability for the project activities. Some contractors may want a daily bar chart (sorted to show only their activities), which they can use to show status based on their daily performance. This chart will not show float availability.

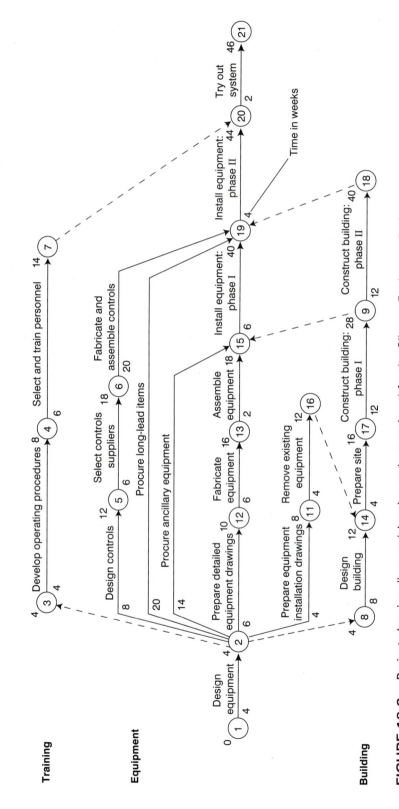

Training

Equipment

Building

FIGURE 10.6 Project planning diagram (showing early start times) for the Gilmore Equipment Project

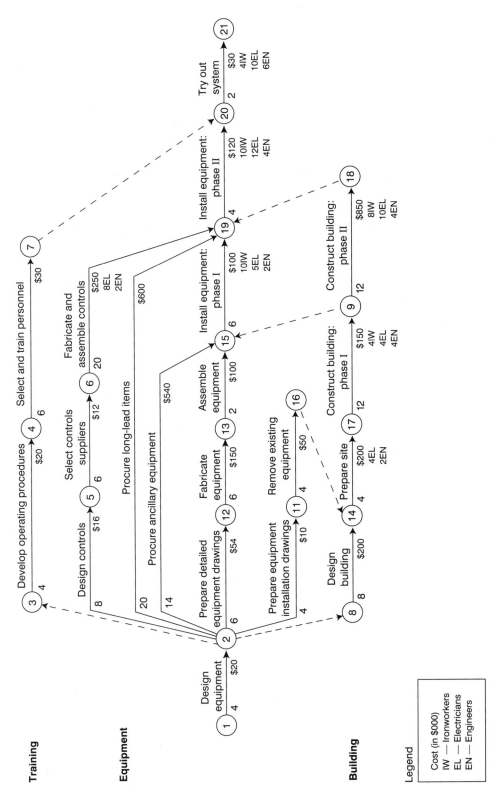

Training

Equipment

Building

Legend

Cost (in $000)
IW — Ironworkers
EL — Electricians
EN — Engineers

FIGURE 10.7 (Master) project planning diagram for the Gilmore Equipment Project

Start Node	End Node	Duration (Days)	Description	Resources (Daily Requirements)		
				IW	EL	EN
Start	1	—				
1	2	20	EQB Design equipment			
2	3	—	Dummy			
2	11	20	EQC Preparation of equipment installation drawings			
2	15	70	EQE Procure ancillary equipment			
2	12	30	EQF Prepare detailed equipment drawings			
2	8	—	Dummy			
2	5	40	EQI Design controls			
2	19	100	EQD Procure long-lead items			
8	14	40	BUB Design building			2
14	17	20	BUD Prepare site		4	4
17	9	60	BUE Construct phase I building	4	4	4
9	18	60	BUF Construct phase II building	8	10	
16	14	—	Dummy			
9	15	—	Dummy			
18	19	—	Dummy			
5	6	30	EQJ Select controls suppliers			
6	19	100	EQK Fabricate and assemble controls		8	2
3	4	20	TRB Develop operating procedures			
4	7	30	TRC Select and train personnel			
7	20	—	Dummy			
12	13	30	EQG Fabricate equipment			
13	15	10	EGH Assemble equipment			
15	19	30	EQL Install equipment phase I	10	5	2
19	20	20	EQM Install equipment phase II	10	12	4
11	16	20	BUC Remove existing equipment			
20	21	10	EQN Try out system	4	10	6
21	Finish	—				

IW — Ironworkers
EL — Electricians
EN — Engineers

FIGURE 10.8 Worksheet for the Gilmore Equipment Project

PROJECT MANAGEMENT   GILMORE EQUIPMENT PROJECT   FARMINGTON, MICHIGAN
      REPORT TYPE :STANDARD LISTING   PRINTING SEQUENCE :Node/Activity Sequence
                                              SELECTION CRITERIA :ALL
      PLAN I.D.   :GEP   VERSION 3   TIME NOW DATE   : 8/NOV/95

ACTIVITY DESCRIPTION	EARLIEST START	EARLIEST FINISH	LATEST START	LATEST FINISH	DURATION	FLOAT
8- 14 BUB DESIGN BUILDING	6/DEC/95	30/JAN/96	6/DEC/95	30/JAN/96	40	0 *
11- 16 BUC REMOVE EXISTING EQUIP	3/JAN/96	30/JAN/96	3/JAN/96	30/JAN/96	20	0 *
14- 17 BUD PREPARE SITE	31/JAN/96	27/FEB/96	31/JAN/96	27/FEB/96	20	0 *
17- 9 BUE CONST BLDG PHASE I	28/FEB/96	21/MAY/96	28/FEB/96	21/MAY/96	60	0 *
9- 18 BUF CONST BLDG PHASE II	22/MAY/96	13/AUG/96	22/MAY/96	13/AUG/96	60	0 *
1- 2 EQB DESIGN EQUIPMENT	8/NOV/95	5/DEC/95	8/NOV/95	5/DEC/95	20	0 *
2- 11 EQC PREP EQUIP INSTAL DWGS	6/DEC/95	2/JAN/96	6/DEC/95	2/JAN/96	20	0 *
2- 19 EQD PROCURE LONG-LEAD ITEMS	6/DEC/95	23/APR/96	27/MAR/96	13/AUG/96	100	80
2- 15 EQE PROCURE ANCILLARY EQUIP	6/DEC/95	9/APR/96	28/FEB/96	2/JUL/96	90	60
2- 12 EQF PREPARE DETAIL EQUIP DWG	6/DEC/95	16/JAN/96	27/MAR/96	7/MAY/96	30	80
12- 13 EQG FABRICATE EQUIPMENT	17/JAN/96	27/FEB/96	8/MAY/96	18/JUN/96	30	80
13- 15 EQH ASSEMBLE EQUIPMENT	28/FEB/96	12/MAR/96	19/JUN/96	2/JUL/96	10	80
2- 5 EQI DESIGN CONTROLS	6/DEC/95	30/JAN/96	20/DEC/95	13/FEB/96	40	10
5- 6 EQJ SELECT CONTROLS SUPPLIERS	31/JAN/96	12/MAR/96	14/FEB/96	26/MAR/96	30	10
6- 19 EQK FAB AND ASSEMBLE CONTROLS	13/MAR/96	30/JUL/96	27/MAR/96	13/AUG/96	100	10
15- 19 EQL INSTALL EQUIP PHASE I	10/APR/96	21/MAY/96	3/JUL/96	13/AUG/96	30	60
19- 20 EQM INSTALL EQUIP PHASE II	14/AUG/96	10/SEP/96	14/AUG/96	10/SEP/96	20	0 *
20- 21 EQN TRYOUT SYSTEM	11/SEP/96	24/SEP/96	11/SEP/96	24/SEP/96	10	0 *
3- 4 TRB DEVELOP OPERATING PROCEDURES	6/DEC/95	2/JAN/96	3/JUL/96	30/JUL/96	20	150
4- 7 TRC SELECT AND TRAIN PERSONNEL	3/JAN/96	13/FEB/96	31/JUL/96	10/SEP/96	30	150

PROJECT MANAGEMENT   GILMORE EQUIPMENT PROJECT   FARMINGTON, MICHIGAN
      REPORT TYPE :COMPRESSED PERIOD BARCHART   PRINTING SEQUENCE :Node/Activity Sequence
                                          SELECTION CRITERIA :ALL
      PLAN I.D.   :GEP   VERSION 3   TIME NOW DATE   : 8/NOV/95

```
=====================================1995=======1996=====================================
                                   !8  !4  !1  !5  !4  !1   !6  !3  !1  !5  !2  !7  !4  !
PERIOD COMMENCING DATE
MONTH                              !NOV !DEC !JAN !FEB !MAR !APR !MAY !JUN !JUL !AUG !SEP !OCT !NOV !
PERIOD COMMENCING TIME UNIT        !6  !24  !44  !69  !89  !109  !134 !154 !174 !199 !219 !244 !264 !
=========================================================================================
 8- 14 BUB DESIGN BUILDING          !  ! CCCC!CCCCCC!   !   !   !    !   !   !   !   !   !   !
11- 16 BUC REMOVE EXISTING EQUIP    !  !  ! CCCCC!   !   !   !    !   !   !   !   !   !   !
14- 17 BUD PREPARE SITE             !  !  !  !CCCCC!   !   !    !   !   !   !   !   !   !
17-  9 BUE CONST BLDG PHASE I       !  !  !  !   !CCCCC!CCCCCCC!CCC !   !   !   !   !   !   !

 9- 18 BUF CONST BLDG PHASE II      !  !  !  !   !   !   !CC !CCCCC!CCCCCC!CC !   !   !   !
 1-  2 EQB DESIGN EQUIPMENT         !CCCC!C !   !   !   !   !    !   !   !   !   !   !   !
 2- 11 EQC PREP EQUIP INSTAL DWGS   !  ! CCCC!C  !   !   !   !    !   !   !   !   !   !   !
 2- 19 EQD PROCURE LONG-LEAD ITEMS  !  ! ====!======!=====!=====!===== !   !   !   !   !   !   !

 2- 15 EQE PROCURE ANCILLARY EQUIP  !  ! ====!======!=====!=====!======!=== !   !   !   !   !   !
 2- 12 EQF PREPARE DETAIL EQUIP DWG !  ! ====!==== !   !   !   !    !   !   !   !   !   !   !
12- 13 EQG FABRICATE EQUIPMENT      !  !  !  ===!====== !   !   !    !   !   !   !   !   !   !
13- 15 EQH ASSEMBLE EQUIPMENT       !  !  !  ! !=== !   !   !    !   !   !   !   !   !   !

 2-  5 EQI DESIGN CONTROLS          !  ! ====!======!   !   !   !    !   !   !   !   !   !   !
 5-  6 EQJ SELECT CONTROLS SUPPLIERS!  !  !  !=====!=== !   !    !   !   !   !   !   !   !
 6- 19 EQK FAB AND ASSEMBLE CONTROLS!  !  !  !   !  ===!=======!=====!=====!======! !   !   !   !
15- 19 EQL INSTALL EQUIP PHASE I    !  !  !  !   !  =====!=== !    !   !   !   !   !   !   !

19- 20 EQM INSTALL EQUIP PHASE II   !  !  !  !   !   !   !    !   !   !CCC!CC !   !   !   !
20- 21 EQN TRYOUT SYSTEM            !  !  !  !   !   !   !    !   !   !   ! CCC !   !   !
 3-  4 TRB DEVELOP OPERATING PROCEDUR!  ! ====!= !   !   !   !    !   !   !   !   !   !   !
 4-  7 TRC SELECT AND TRAIN PERSONNEL!  !  ! =====!=== !   !   !    !   !   !   !   !   !   !
=========================================================================================
```

Barchart Key:-   CCC :Critical Activities   === :Non Critical Activities   NNN :Activity with neg float

FIGURE 10.9   (a) Tabulated schedule for the Gilmore Equipment Project; (b) critical activities bar chart for the Gilmore Equipment Project

MEASURING PROJECT PERFORMANCE

To control the project effectively, M&A and all of the participants need to communicate frequently and constantly. While monthly progress reports are useful, the participants, especially in construction and installation, must have more direct knowledge of progress in shorter time. Project control, which continues over the life of a project, involves monitoring, assessing, taking action on problems, and communication.

M&A will conduct early-morning meetings daily with supplier representatives, contractors, and owner representatives, discussing such items as late equipment deliveries, work interferences among the contractors, and scheduling of interrelated work. Figure 10.10 shows a checklist of potential causes of timing problems. This is used as a guide to get to the root of the problem.

At regular weekly meetings key equipment supplier personnel, contractor supervisory personnel, and engineering representatives of the design/build firm will review the work to be done in the next few weeks (and months in the case of

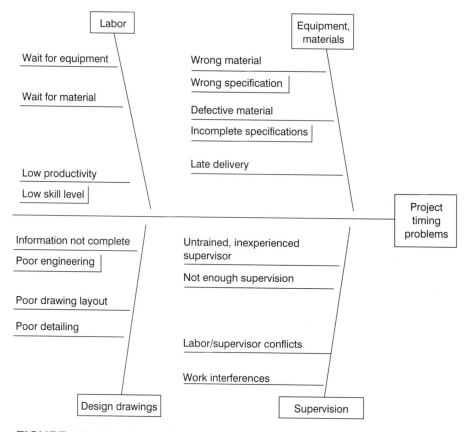

FIGURE 10.10   Cause/effect diagram of potential timing problems

strategic items). As facilitator at the meetings, M&A will emphasize project timing awareness as a major priority. Typical discussions may concern late equipment deliveries, and the possibility of shipping equipment piecemeal to keep the job running and on target. In every meeting M&A will underscore, in addition to timing, two of the most important issues:

- **safety** in the workplace—properly designed barricades surrounding excavations, strategically placed fire extinguishers, designated hard hat areas, regularly scheduled safety talks, and so on
- **quality** in the workplace—materials, equipment, and installation performance

M&A assigns a person to document the meeting discussions in minutes and to distribute the minutes promptly to all attendees. These become valuable as a guide for discussions at the next meeting. Participants use them as references to report on the action they have taken on assignments received at the previous meeting.

Whether discussing project performance or preparing status reports, M&A uses the same theme as an outline:

1. **Monitor** the present status of each of the project items now under way or soon to be started.
2. **Assess**—identify the critical items and discuss their problems.
3. **Resolve** by making corrections and adjustments to bring the project back to plan.
4. **Communicate** through constant contact with project participants; encourage direct contact among project participants.

Figure 10.11 shows a typical communication document format used to record the daily equipment installation contractor activities. This type of record is useful to M&A, as well as to the contractors, as reference information. It will help settle differences concerning work items and disputed extras. Most contractors and subcontractors are credible and conduct their business with integrity; however, there will be differences of opinions on questions of past work. Well-kept records will narrow these differences. Thorough understanding of owner and contractor positions can prevent difficult situations. Their common goals for a successful project should include completing it on time, within budget, and at a high level of quality.

Preparing Project Status Reports

The summary status report, prepared monthly for the owner and various levels of M&A management, essentially capsulizes the meeting minutes. M&A will not include much detail, but will instead focus on important concerns, using project summary schedules and bar charts for exhibits.

Daily Report on Equipment Installation

_____ Plant

Project and Item Number _____

Date _____

Location _____

Temperature _____

Purchase Order _____

Weather _____

General Contractor _____

Daily Personnel Breakdown

Contractor and Sub-Contractors	Trades												Total
	Brick-layers	Carp.	Elect.	Iron-workers	Lab.	Mill-wrights	Pipe-fitters	Riggers	Sheet metal				

Work Progress (equipment or process being installed, % complete, specific problem, etc.) _____

_____

_____

_____

_____

_____

_____

_____

_____

_____

_____

_____

_____

_____

Authorized Additional Work (notice to proceed, personnel, status, etc.) _____

_____

_____

_____

Equipment Received (manufacturer, scheduled delivery date, condition at arrival, etc.) _____

_____

_____

_____ Signed _____

FIGURE 10.11   Daily report on equipment installation for the Gilmore Equipment Project

The schedules used in the summary status report are the latest start and finish dates with pertinent float times for the highlighted project items. For example, the delivery date of an equipment item is drifting from the original schedule; however, the amount of float available for this item is adequate to handle that drift. The report will conclude this equipment item is no cause for concern at this time. The report notes that it is flagged, however, for signs of any adverse trend in the weeks ahead. M&A will be prepared to take corrective measures when and if the project completion date appears to be jeopardized.

On the other hand, the progress schedule report, sent to contractors and suppliers, reflects the early start and finish schedule and does not mention available float. Therefore, all of those directly engaged in executing the project need to adhere strictly to the original schedule. M&H informs them it cannot tolerate any drifts, which will directly affect the project timing. This tone may sound harsh, but project control requires this type of action for successful completion. There are exceptions, however, allowed only after a thorough investigation of the validity of the request.

Status reports include these documents: cover letter, executive highlights, and summary progress schedule, which contains a bar chart schedule and accompanying commentary.

Let's say we're interested in the status of the Gilmore Equipment Project at the end of March, 1996. After monitoring the project activities with the suppliers and contractors, M&A lists the significant project activities that have not been following the schedule:

Activity	Weeks Behind Schedule
Construct building phase I	2
Fabricate and assemble controls	2
Assemble equipment	6

In assessing the effect of these behind-schedule activities on the project duration, M&A will input its revised durations to generate a new schedule. The new schedule shows the *Try out system* finish date (also the project completion date) is October 8, 1996, two weeks later than the scheduled date, September 24, 1996.

Since the activities listed above directly or indirectly affect the schedule, M&A uses the revised schedule to record the new float times:

Activity	Float (Weeks)	
	Original	Revised
Construct building phase I	0	0
Fabricate and assemble controls	2	2
Assemble equipment	16	12

The activity, *Construct building phase I,* with 0 float, has pushed project timing back two weeks. To recover these two weeks and get back to the original schedule, M&A will now discuss with the phase I building contractors how to subtract two weeks from their remaining work schedule. If the phase I contractors are unable to expedite their work, M&A will explore possible strategies to reduce the phase II building schedule. Because it has exposed the problem early, M&A can analyze various options to bring the project back on schedule.

All the information that M&A has gathered will be used to prepare the regular monthly status report.

Cover Letter.   The cover letter briefly describes the overall project status. M&A will use the cover letter to (1) summarize the progress of the project to date, (2) show anticipated completion dates of major events, (3) recap the status of critical items, and (4) resolve any potential concerns on critical items.

In the April 1, 1996 status report, M&A states in the opening paragraph of the cover letter that the project may be delayed by as much as two weeks. This statement should get readers' immediate attention. A brief explanation follows, with an account of how this concern may be resolved. The letter continues to describe the status of other pertinent project activities.

Figure 10.12 is the cover letter for the April 1, 1996 report. The reader, who is interested in knowing more of the status, will review the attachments, executive highlights, and summary progress schedule. M&A designs the cover letter so the reader will want to examine the executive highlights.

Executive Highlights.   M&A adds an executive highlights page as an attachment to the cover letter. It includes important incidents of the project that will influence the status, listing each with a bullet. These conditions are brief—a sentence or two. If M&A prepares a short progress status report, it can include the highlights in the body of the cover letter.

The executive highlights for April 1, 1996 (shown in Figure 10.13) amplify the information included in the cover letter. The document includes information on potential building construction delays and other concerns (equipment deliveries) with an explanation of how the problems will be handled.

Summary Progress Schedule.   An optional attachment to the cover letter is a summary progress schedule. It includes

- a bar chart illustrating the project status, displaying changes to the planned completion date and actual status of the major project activities
- a brief commentary of the status, highlighting any special concerns

A summary progress schedule gives Gilmore Industries and M&A the ability to review the overall project progress on one page. This type of schedule summarizes all the countless details associated with day-to-day activities. Using the management-by-exception technique of concentrating attention on critical items, the reader soon comprehends the concerns and can participate in solutions.

M & A ENGINEERING
FARMINGTON, MI

April 1, 1996

Gilmore Industries
Farmington, MI

Subject: Status of Gilmore Equipment Project

Based on our latest status review, we anticipate completion of the project on October 8, 1996, when launching the product can begin. This is two weeks later than the original plan.

Because of the exceptionally heavy and frequent rainfall experienced in March, the phase I construction has been curtailed. The contractor estimates that the work is behind two weeks, and the time will be difficult to recover. As the work in phase I construction is critical, the project completion date has been directly affected.

Equipment suppliers have reported encountering delays on several equipment items and selected controls; however, they are not affecting the overall project. Notifications have been sent to the suppliers that they cannot report any more delays without severely affecting the progress of the project.

The training coordinator has reported the completion of the training manuals expected by April 15 (with the exception of the standards to be included in the quality assurance manuals). She also reports that all of the personnel for the new facility are now employed, and their training will start within a week or two after completion of the training manuals.

We have asked the phase II construction contractors to review all of their critical work activities with the objective of reducing their operations by two weeks. It is anticipated that they will report back that the timing of several items can be shortened, but at some additional costs. Our reviews with them will center on minimizing these costs.

Please refer to the *Highlights* and the *Summary progress schedule* that are attached for further details. Please let us know if you have any comments or questions.

Sincerely yours,

William A. Rhodes
Project Manager

FIGURE 10.12   Status report cover letter for the Gilmore Equipment Project

---

Gilmore Equipment Project

EXECUTIVE HIGHLIGHTS
APRIL 1, 1996

BUILDING
- Phase I building construction contractors experiencing delays due to abnormal weather conditions—rainfall in March 75% over normal. Building construction behind about two weeks.

EQUIPMENT
- Electronic controls supplier reports that quality problems have delayed parts shipments. Requests that controls deliveries be extended from July 30 to August 13. (Adequate float—does not affect schedule.)
- Equipment assembly company working on key equipment meeting with unauthorized work stoppages. Expects settlement within week, but will not be back to normal operations for six weeks. Will now deliver April 28. (Adequate float – does not affect schedule.)

TRAINING
- Completed enrollment of personnel for new facility on March 15.
- Training coordinator reports that all training manuals except quality assurance volumes will be available April 15. Classes expect to start June 1.

M&A Engineering
Farmington, MI

FIGURE 10.13   Status report: Executive highlights for the Gilmore Equipment Project

On many projects, a progress status report will include the status concerning project costs as well as project timing. Where preserving confidentiality of costs is necessary, a separate cost report is prepared for limited viewing. The following section describes the type of cost reports that would be submitted.

MEASURING COST PERFORMANCE

Managing costs and managing time require the same efforts: planning, scheduling, and controlling.

Planning costs for a project forms the basis for the authorized (or budgeted) costs for the project. The budget is the total sum of the costs needed to complete each activity. In most cases, these costs reflect estimates from knowledgeable

persons; however, uncertainty can lessen their accuracy, which may have significant consequences in the long run. On most projects, meeting the budget takes
precedence over keeping the job on time when measuring performance.

Those formulating budgets for certain types of projects can use computerized
estimating programs. Estimating software is becoming more prevalent, especially
for construction and facility projects. These programs contain estimating information based on historical data, as well as for various types of projects. The
information may be refined to show particular situations. In addition, computer
programs now being introduced assist in the "quantity takeoffs" (measuring
material and products) from design drawings. This type of estimating has a high
degree of credibility, since it calculates based on completed design and construction drawings.

Proposals from contractors and suppliers offer additional estimating sources
of material, labor, and equipment costs. Using these sources, however, may pose
problems in timing. These estimates are usually not available when the budget
needs to be completed. Estimates are needed during the planning phase, well
ahead of any authorization to proceed. In some cases of no-tolerance estimates,
some expenditures may be authorized before budget approval in an attempt to
pinpoint important costs. See Figure 10.7 for the cost estimates prepared for the
Gilmore Equipment Project.

A cost schedule will answer these questions: (1) when will money be needed
to make payments for labor, material, and other expenses over the duration of
the project, and (2) how much money will these expenses require?

Computer programs will generate incurred costs over any desired period—
daily, weekly, or monthly. Using a worksheet with its tabulation format will
expedite inputting by clarifying the data required for this effort. The worksheet
contains data needed for computer input, including each activity's identification,
description, and total costs to complete. To conform to the computer program
being used, the worksheet also shows the unit costs of each activity. Figure 10.14
shows a completed worksheet used to input cost data for the Gilmore Equipment
Project.

The computer program used here treats costs as another resource—
like labor and personnel—and generates a histogram to reflect the costs
expended throughout the schedule. (The computer can produce cost reports both
in graphic and in tabular form.) M&A uses these printouts to analyze the budget,
its schedule, and its ability to meet all requirements. Figures 10.15 and 10.16,
which show the graphic spending schedule for the Gilmore Project, will help
M&A answer these questions categorically: January and February, 1996 are the
highest spending months, and the daily spending range in that period is
$28,000–$32,000.

The cost schedule reports help in forecasting funds needed to operate this
project in the future. M&A, with the aid of the computer, can quickly analyze any
changes in the project that may influence future costs. Computerized cost schedules are invaluable for what-if exercises to resolve financial problems.

Start Node	End Node	Duration (Days)	Description	Costs in Dollars		
				Total Costs	Daily Costs	
					TC	$C
Start	1	—				
1	2	20	EQB Design equipment	20,000	1,000	1,000
2	3	—	Dummy			
2	11	20	EQC Preparation of equipment installation drawings	10,000	500	500
2	15	70	EQE Procure ancillary equipment	540,000	7,714	7,714
2	12	30	EQF Prepare detailed equipment drawings	54,000	1,800	1,800
2	8	—	Dummy			
2	5	40	EQI Design controls	16,000	400	400
2	19	100	EQD Procure long-lead items	600,000	6,000	6,000
8	14	40	BUB Design building	200,000	5,000	5,000
14	17	20	BUD Prepare site	200,000	10,000	10,000
17	9	60	BUE Construct phase I building	150,000	2,500	2,500
9	18	60	BUF Construct phase II building	850,000	14,167	14,167
16	14	—	Dummy			
9	15	—	Dummy			
18	19	—	Dummy			
5	6	30	EQJ Select controls suppliers	12,000	400	400
6	19	100	EQK Fabricate and assemble controls	250,000	2,500	2,500
3	4	20	TRB Develop operating procedures	20,000	1,000	1,000
4	7	30	TRC Select and train personnel	30,000	1,000	1,000
7	20	—	Dummy			
12	13	30	EQG Fabricate equipment	150,000	5,000	5,000
13	15	10	EGH Assemble equipment	100,000	10,000	10,000
15	19	30	EQL Install equipment phase I	100,000	3,333	3,333
19	20	20	EQM Install equipment phase II	120,000	6,000	6,000
11	16	20	BUC Remove existing equipment	50,000	2,500	2,500
20	21	10	EQN Try out system	30,000	3,000	3,000
21	Finish	—				

TC — Project Activity Costs
$C — Accumulative Cost

FIGURE 10.14   Cost worksheet for the Gilmore Equipment Project

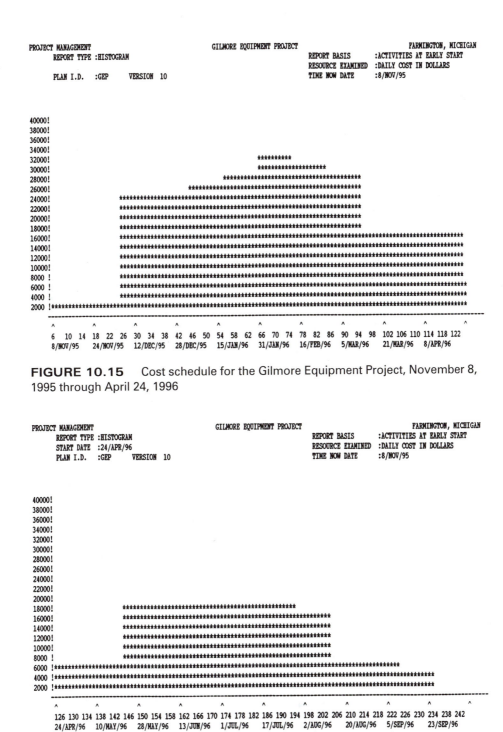

PROJECT MANAGEMENT   GILMORE EQUIPMENT PROJECT   FARMINGTON, MICHIGAN
    REPORT TYPE :HISTOGRAM   REPORT BASIS   :ACTIVITIES AT EARLY START
                                                                           RESOURCE EXAMINED :DAILY COST IN DOLLARS
    PLAN I.D. :GEP   VERSION 10   TIME NOW DATE   :8/NOV/95

FIGURE 10.15   Cost schedule for the Gilmore Equipment Project, November 8, 1995 through April 24, 1996

PROJECT MANAGEMENT   GILMORE EQUIPMENT PROJECT   FARMINGTON, MICHIGAN
    REPORT TYPE :HISTOGRAM   REPORT BASIS   :ACTIVITIES AT EARLY START
    START DATE :24/APR/96   RESOURCE EXAMINED :DAILY COST IN DOLLARS
    PLAN I.D. :GEP   VERSION 10   TIME NOW DATE   :8/NOV/95

FIGURE 10.16   Cost schedule for the Gilmore Equipment Project, April 24, 1996 through September 24, 1996

M&A employs cost control in the same manner as it managed the project timing schedule. Using an approach known as the **indicated cost method**, the company can detect at an early date any potential overspending or cost overruns.

The indicated cost report, an early warning system, alerts designated participants of the status of project spending. It determines the ability of the costs of any project activity to fall within the authorized budget. This report is prepared periodically, usually coinciding with other status reports, and reviews and evaluates in detail project spending. The contents include: (1) a comparison of the project spending with budget allowances, (2) the amount of additional spending forecast, (3) identification of the activities with overruns (overspending their authorized costs) that have contributed to the forecasted additional spending, (4) an analysis of the overruns, and (5) recommendations to reduce overspending.

The April 1, 1996 indicated cost report for the Gilmore Project (Figure 10.17) shows a variance of $198,000, or 5.6 percent of the total budget. This concern and the possibility that the trend of overspending could continue prompts M&A to direct a cost review of the total project. The review is necessary because the variance is so high on a project that is only half completed. The overspending trend is alarming. The cost report indicates the major contributors to the overrun: procurement of the long-lead equipment items and the ancillary equipment items. These two items alone account for 60 percent of the total overrun, $120,000.

Meetings with the engineers and suppliers reveal that costly design changes were authorized as the detail drawings were being completed. Gilmore Industries had agreed that these changes expanded the equipment capability, and the additional costs would be repaid in higher productivity. At the time, the company rationalized that it could find a way to pay for the changes without adding to the costs of the overall project. In this case, it may be right: To pay for the extra costs, M&A has recommended that suppliers reduce the initial spare parts complement for this equipment. The spare parts inventory, used for equipment maintenance, will be held to a minimum. Gilmore is relying on equipment suppliers, who have assured the company that their high-quality work will result in minimal equipment downtime.

As shown in the indicated cost highlights (Figure 10.18), M&A has specified additional recommendations for bringing the project within budget. The firm has pinpointed future construction and equipment installation items for possible cost reduction areas.

If it does not breach financial confidentiality, M&A will include the indicated cost outcome and indicated cost highlights reports with the project status reports. Otherwise, distribution of the cost information will be limited to Gilmore Industries and M&A management.

REPORTING DATE: 4/1/96 STARTING DATE: 11/8/95 COMPLETION DATE: 9/24/96

Project Activity	Authorized (Budgeted)	Committed to Date	Future Commitments	Indicated Outcome	Variance (over) or under	% Variance (over) or under
Design equipment	20	20		20		
Prepare equipment installation drawings	10	10		10		
Procure ancillary equipment	540	530	50	580	(40)	(7.4)
Prepare detailed equipment drawings	54	54		54		
Design controls	16	16		16		
Procure long-lead items	600	600	80	680	(80)	(13.3)
Design building	200	240		240	(40)	(20.0)
Prepare site	200	215		215	(15)	(7.5)
Construct building phase I	150	130	40	170	(20)	(13.3)
Construct building phase II	850		850	850		
Select controls suppliers	12	12		12		
Fabricate and assemble controls	250	175	75	250		
Develop operating procedures	20	18		18	2	10.0
Select and train personnel	30		25	25	5	16.7
Fabricate equipment	150	160		160	(10)	(6.7)
Assemble equipment	100	100		100		
Install equipment phase I	100		100	100		
Install equipment phase II	120		120	120		
Remove existing equipment	50	50		50		
Try out system	30	30		30		
	3,502	2,360	1,340	3,700	(198)	(5.6)

Note: Costs in $000
Project completion: 44.5%

FIGURE 10.17 April 1, 1996 indicated cost outcome for the Gilmore Equipment Project

Gilmore Equipment Project
INDICATED COST HIGHLIGHTS
APRIL 1, 1996

- Project (authorized) budget : $3,502,000
- Project commitments to date : $2,360,000
  % of budget   67%
- Cost variance : $198,000 (5.6%) over budget
- Completion of project (%) : 44.5% complete
- Outstanding activities (not completed) with high overruns:

		$(000)
a.	Procure ancillary equipment	(40)
b.	Procure long-lead items	(80)
c.	Construct building phase I	(20)
d.	Fabricate equipment	(10)
	Total	$(150)

- Recommendations to reduce overrun:

		REDUCTION $(000)	RECOMMENDATION
a.	Procure ancillary equipment	20	Reduce selected spare parts
b.	Procure long-lead equipment	50	Reduce selected spare parts
c.	Install phase I, II equipment	40	Employ plant forces to install selected nonproduction (maintenance, material handling) items
d.	Construct phase II	85	Employ plant forces to complete nonessential items
		$195	

Results of recommendations to be reported at next status report.

FIGURE 10.18   Indicated cost highlights for the Gilmore Equipment Project

CONCLUSION

A major responsibility of M&A is controlling the time and costs (by measuring performance) over the span of the Gilmore Equipment Project. Acceptable performance means, to a large extent, satisfactorily meeting the time and cost objectives set by Gilmore Industries.

Measuring project performance consists of four steps:

1. Monitor—collect information on the project status through one-on-one contacts, meetings, and reports.

2. Assess—analyze the status information, including using computer reports generated from the status data, to measure performance.
3. Resolve—take action on critical items that are adversely affecting the completion date objectives. Observe and take note of any potential improvements to the schedule.
4. Communicate—inform participants about the project status through memos, meetings, and written reports. Highlight the plans to resolve problems.

Adequately measuring the performance of a project needs a base plan and schedule. The time spent in thorough preparation of the planning diagram and the computerized schedule is worthwhile when comparing the present situation with the original objectives.

Examining computer printouts highlights the critical and behind-schedule activities. A useful status report will (1) highlight activities, both critical and non-critical, that are behind schedule, (2) emphasize evaluation of the critical items, and (3) show resolutions to restoring satisfactory performance.

This approach is designed to help in early detection of concerns to allow enough time to resolve problems before they become serious situations. The effort of preparing a thorough status review that forecasts potential problems six months ahead has its rewards.

Communication is the key to successful projects. One-on-one contacts, meetings, memos, and reports combine to inform all participants of their role in bringing the project within budget and on time.

Controlling costs is not unlike controlling timing of a project. Cost control uses an early warning system, the indicated cost outcome method, which targets overspending situations and offers recommendations to correct and maintain satisfactory project performance.

Exercises

1. M&A needs to revise the diagram in Figure 10.5 that shows the project activities and their relationships because of some new decisions:
   • Gilmore Industries wants to place the personnel who will be operating and maintaining the new facility in the plant as soon as they are trained (i.e., when *Install Equipment Phase I* starts instead of when *Install Equipment Phase II* ends).
   • When the *Design* activity is completed, both *Remove existing equipment* and *Prepare equipment installation drawings* can start and be done concurrently.
      a. Revise the planning diagram to reflect the new positions of *Remove existing activity* and *Prepare equipment installation drawings* and show when the trained personnel will be placed in the plant.
      b. Do the above changes affect the completion date? If so, what is the new date? (Use the manual method for calculating the earliest start times and locate them on the planning diagram.)

   c. Input the node changes and generate a new schedule report for
the Gilmore Equipment Project. Complete the float comparison
tabulation shown below:

Project Activity	Original Schedule	Revised Schedule

2. After meeting with some of the potential equipment suppliers soon after
completing the original planning diagram, M&A decided to change the
duration of several activities:

Activity	Change
Procure long-lead items	From 20 to 18 weeks
Assemble equipment	From 2 to 4 weeks

   Add the above changes to the revised plan and schedule in Exercise 1 and
answer the following questions: What is the new completion date? What
activities now have 0–2 float? Since the original project performance now
may be in jeopardy with the new dates, what changes could be made to
return to the initial schedule?

3. Gilmore Industries instructs M&A to start the Gilmore Equipment Project
on January 3, 1996 instead of November 8, 1995 because of some internal
difficulties that will delay acquisition of the funds needed for this project.
The new objective is to complete the project and start operations about
November 15, 1996. To answer the following questions to determine what
effect this will have on the schedule developed in Exercise 2, you will need
first to change the start and finish input data to reflect the new start and
completion dates, then produce a new computerized schedule.

   a. List the most significant schedule changes (e.g., activities with 0 float
and negative float, if any).

   b. If there is negative float, what activities would you suggest changing to
restore the November 15 completion date?

Product Development Project: Create Milestones

A typical product development project involves these major functions: marketing, product design/development, quality, engineering/manufacturing, and administrative/training. Personnel from these departments will work together through a series of steps, including product concepts, design, development and test; manufacturing; marketing, sales, and distribution.

Product development planning originates with a clear understanding of the objectives and the overall goals. Top management defines the major objectives for making the new product. These are usually accompanied by general guidelines and procedures for the product planners to follow. Product development programs with distinct phases work well for applying project management principles. Guidelines include the preparation of training programs, ranging from highly technical subjects to project management principles, to self-improvement programs for people assigned to the project.

Those assigned to this project will report to a program leader. Although their new role is separate from their permanent position, and is essentially a temporary assignment, they must give full attention to their new assignment to help ensure the project's success.

BACKGROUND

Roy Dale, President of Dale, Inc., sent the letter shown in Figure 11.1 to the managers of the major departments.

"Team Under One Roof"

The Dale memo conveys an important lesson learned from project management: early involvement of people from departments germane to the project. This group forms the nucleus of the "team under one roof" principle in which early problems are found and fixed collectively. Eliminating changes that would inevitably occur later during fabrication and production avoids a potentially time-consuming and costly situation.

Teaming up people from various disciplines minimizes paperwork and meetings and maximizes informal discussions. Informal decision making between parallel teams also enhances the team's "ownership" of the decision and ultimately the product. Team members "own" (or buy into) a decision more readily than when they communicate from one area to another, one building to another, or possibly from one city to another. (One manager's statement: "Someone could ask me a question at the coffee pot, and I could make a decision in 10 seconds that might have taken three weeks if it had to go through the various departments.")

This type of strategy permits simultaneous (also called concurrent) engineering, which is simply bringing involved personnel together every day to proceed with their assignments, and work more smoothly. Real-time problem solving and decision making on a collective basis become matter-of-fact under this type of an environment.

Extending the simultaneous engineering idea is computer-aided engineering (CAE), which permits the engineers to design, test, and analyze new components simultaneously before building the prototype. This saves time and money and improves the overall quality of the component. CAE, with centralized work stations, enhances the under-one-roof concept by allowing all of the work to be done in one location rather than in several different areas. The time saved can be used to examine more alternatives.

Real-time project management enhances the under-one-roof concept by allowing team members to know exactly where they are in the project (its status) at any given time and to deal immediately with any barriers to achieving the project goals. Real-time project management begins with setting goals and preparing a project planning diagram.

CONCEPT TO MARKET: PRODUCT PLANNING CYCLE

The concept-to-market (CTM) cycle with the team under one roof is aimed at bringing the product to market faster, more efficiently, and with higher quality.

August 2, 1996

DALE INDUSTRIES

TO: MESSRS. K.J. Kunkle   Sales and Marketing
           W.T. Lindt   Product Design
           J.M. Loile   Engineering and Manufacturing
           W.P. Port   Quality
           G.R. Schuette   Administration/Training

Subject: Concept to Market (CTM)--Project Management
           Workshop

As you know, we are planning to develop a new product that will improve our market penetration. To accomplish this, we need to bring our product to market faster, improve our present quality practices, and enhance customer satisfaction. We will achieve this by improving a number of our present operations and by using appropriate project management principles. Using the project management precept will help control our planning costs, plan our resources, and deal proactively with the program to resolve issues before timing becomes critical.

To "quick start" the program, please support this effort by promptly delegating two people from your department (preferably those who will be assigned to this project) to attend a three-day workshop on September 11, 12, and 13. The purpose of the workshop is to form nucleus project teams and to give background information on the project management techniques. The workshop will also initiate the product development plan that will include costs, responsibilities, timing, and other resources required to support CTM.

An important aspect of the seminar is for the teams to focus on their specific responsibilities in the project as they relate to the main CTM milestones. All of their areas are interrelated and it is important that these interrelationships are defined on the planning model. Of equal importance is setting up the responsibilities for the purpose of accountability.

After this workshop is completed we will have a workable plan to execute effectively this important project. You will be receiving an agenda shortly and a request to list those people who will be attending the workshop.

Roy Dale

FIGURE 11.1   Letter from Roy Dale to department managers

Companies that have used this approach have seen reductions in time, costs, and defects through improved communications, simultaneous engineering, and real-time problem solving and decision making.

Mounting a successful product introduction to meet customers' requirements needs to be done in explicit development stages. Such an approach allows the project to be checked at defined intervals. Measuring progress with major milestones that are established at the start or end of each phase will provide excellent project control. The seminar mentioned in Dale's memo is built around CTM, a phased product planning cycle that points the program in a specific direction. With a systematic program of this nature the team can plan, monitor, and review the program at designated intervals.

Different companies use many versions of product planning cycles, all tailored to the company's structure and to the nature of the product to be marketed. Intense competition among product-oriented firms has resulted in continuous revisions to product development cycles—primarily quality improvements and reductions in cycle time. Automotive firms have been especially active in striving to improve quality and reduce cycle time, using methods of selected Japanese and domestic manufacturers as benchmarks. Ford Motor Co., within a five-year span, has reduced the cycle time on most new models from 5.5 years to 3 years. With several of their latest vehicle programs, using the cycle plan, world class timing (WCT), they want to reduce their cycle time to less than three years.

The CTM cycle plan used by Dale, Inc. includes significant milestones that management considers suitable for the job it has to do. Milestones are events; the starting and/or completion points of each phase are significant milestones for this project. To prepare the entire milestone schedule, the project team will identify the major activities within each phase that are needed to complete that specific phase. When scheduling the project, these dates, known as interim milestones, become important for maintaining control.

In the seminar, the team identified the following major activities for each phase:

Phase I: Program Concepts

- Assemble team
- Define the product
- Feasibility study

Phase II: Product Design

- Product specifications
- Product design
- Quality targets
- Plan and schedule
- Plan resources

Phase III: Product Development

- Complete product designs
- Develop quality plan
- Assess product capability

Phase IV: Product Test

- Develop prototype
- Test product
- Authorize final tooling
- Authorize facilities

Phase V: Production

- Install tooling
- Launch
- Ship product

Figure 11.2 shows the CTM plan, expressed in phases, including the major activities to complete each phase.

The completed plan of action will contain the necessary succession of work breakdown structure (WBS)-type levels of activities needed to complete each major activity of each phase. This is the next step for the project team to complete. The project team will use the basic project management principles taught in the workshops to complete these steps:

1. Prepare the WBS.
2. Construct the planning diagram:
    - Identify interrelationships among the activities.
    - Determine the durations for each activity.

Phase I: Program Concepts

Completing phase I, as quickly as possible, is top priority as the contents of the feasibility study will determine whether the project is a "go" or "no-go." Dale, Inc.'s management review of the feasibility report will determine whether the company should proceed with this project. The project team will prioritize the work associated with initially producing the phase I WBS using the CTM phasing and milestone chart as a guide.

Preparing Phase I WBS. The completed phase I WBS is shown in Figure 11.3. Marketing has the main responsibility of assembling and completing the feasibility report. Market research will assess the market climate, identify the customers' wants and needs, study the competitors' product and pricing schedule, and evaluate the overall market potential for the product. Marketing also needs to determine preliminary distribution and advertising costs and, with the help of the engineering/manufacturing team, manufacturing costs. The marketing team also will estimate preliminary facility costs. Engineering/manufacturing, upon completion of its manufacturing feasibility study, may express need for additional facilities. Preliminary facility costs will be included within the feasibility study.

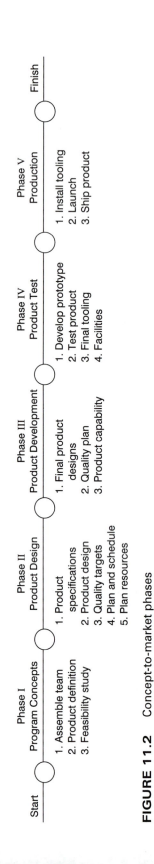

FIGURE 11.2 Concept-to-market phases

Phase I — Program Concepts
1. Assemble team
2. Product definition
3. Feasibility study

Phase II — Product Design
1. Product specifications
2. Product design
3. Quality targets
4. Plan and schedule
5. Plan resources

Phase III — Product Development
1. Final product designs
2. Quality plan
3. Product capability

Phase IV — Product Test
1. Develop prototype
2. Test product
3. Final tooling
4. Facilities

Phase V — Production
1. Install tooling
2. Launch
3. Ship product

Start
Finish

Product Development Project

Phase I: Program Concepts

Marketing	Product Design/ Development	Engineering/ Manufacturing	Quality	Administrative/ Training
1. Identify customer wants, needs	1. Study technology	1. Determine manu-facturing feasibility	1. Analyze potential product failures and causes	1. Procure training materials
2. Evaluate market potential	2. Prepare basic design concepts		2. Assess reliability of the product	2. Select training staff
3. Evaluate competition products and pricing	3. Issue basic pro-gram assumptions			3. Plan training seminars
4. Determine manufac-turing, distribution, and advertising costs				4. Conduct project management seminar
5. Establish pricing schedule				
6. Prepare feasibility study				

FIGURE 11.3 Phase I: Program concepts WBS for the product development project

The quality team also has a meaningful role. Within phase I, the team is responsible for investigating the potential defects that may be experienced with the product assemblies, the product itself, and the causes for these defects. Quality team findings associated with the manufacturing process are incorporated in the engineering/manufacturing study. Within phase I, the quality team's overall assignment is to assess reliability of the proposed product, one of the main topics of the marketing feasibility report.

Administrative/training in phase I initiates its personnel selection and training schedules. Dale management has approved its selection of the project manager (although no formal announcement is made until approval at start of phase II). The project manager will be assisting administrative/training in conducting the project management seminar, which is one of the department's more important responsibilities in phase I.

Constructing the Phase I Planning Diagram.

Figure 11.4 shows the phase I planning diagram for the product development project. No scheduling dates other than the start of phase I (September 11, 1996) and completion of phase I (January 22, 1997) are displayed. To avoid confusion, the planning diagram covers only the activity interrelations and durations. Work relationships among the various functions become conspicuous during this phase. This planning experience gives the teams a sense of the teamwork that the program concepts phase will require for completion as planned. Once the product design/development group has completed the basic design concepts, the product design/development, quality, and engineering/manufacturing teams will be continually interchanging information. Marketing, responsible for preparing the feasibility report, must wait for completed studies by the other groups before arranging the total "package" for presentation to management. The planning diagram plainly shows that all of the studies from the other groups must be completed before marketing can begin assembling the feasibility study package.

Once the logic of the planning diagram is done and the project manager approves activity durations, the manual calculating method can provide a quick check of the longest path (critical path). This information should be available upon completion of the project management seminar, which is 12 weeks into the program. The critical path is 19 weeks—the duration from start to finish of phase I. About December 5, 1996 the project manager will inform Dale, Inc. management, based on the existing information, that it can expect to receive the feasibility report about January 22, 1997. To ensure this date will be met, phase I activities will be closely controlled for the next seven weeks.

Once Dale, Inc. has approved the program, one of the first steps in phase II will be preparing the plan and scheduling each of the phases in the project. The procedures used will be the same as those employed to develop phase I.

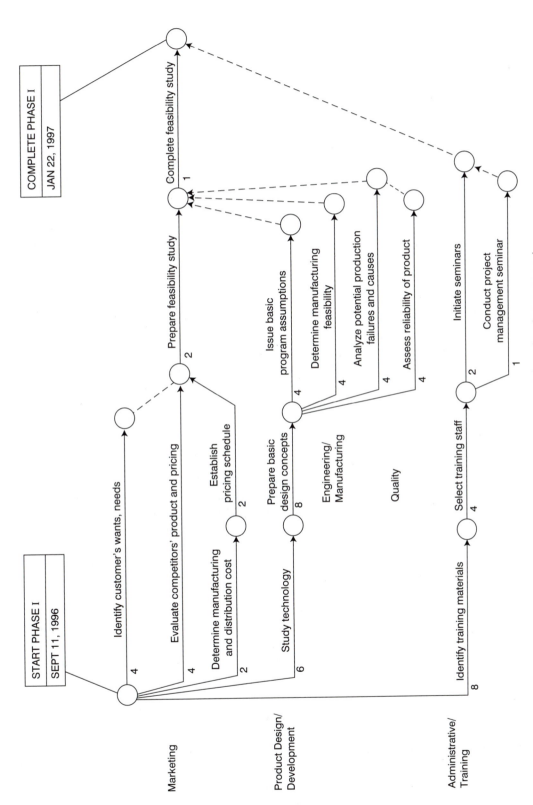

FIGURE 11.4 Phase I program concepts planning diagram for the product development project

SEMINAR WORKSHOPS

Education and training programs have now become an essential part of the standard operating procedures of every successful business. Included among the programs are training seminars and workshops on project management for those participating in new projects. The topics of ensuing workshops go beyond planning and scheduling projects. Among offerings of the training staff are training courses devoted to the specific attributes of the individual participants.

Individual Assessment and Training

Individual training of participants depends on their present qualifications—knowledge, skills, and abilities—for project work. They receive training in planning and organizing skills, leadership, interpersonal skills, communication skills, and technical knowledge.

The September 11–13 seminar includes training in planning and organizing skills. Members of the training team will be observing the participants' ability to think ahead, their handling of unforeseen situations, and the overall organization of their assignments. Individuals who are deficient in any of these attributes will be offered special training courses.

Leadership is an attribute that the project leader obviously needs to possess, but it is also a quality appropriate for all responsible team members. Influencing others toward achieving the goals is the main trait. Motivating, delegating, coaching, and decision making are all necessary interactions between the leader and the team. Training packages cover all of these areas.

Interpersonal skills concern the control of conflict and the aptitude for negotiating. One who possesses this attribute is comfortable among people—at meetings, problem-solving sessions, and day-to-day activities. A typical interpersonal workshop would assess the present interpersonal skills and then develop practice techniques to improve the participant's style.

Communication skills are the listening, speaking, and writing abilities fundamental to the project's success. Team members can effectively transfer thoughts, opinions, ideas, plans, and procedures by sending and receiving suitable communications. Training in speaking skills strives to build self-confidence and reduce anxiety. Workshops to improve writing skills concentrate on preparing clear, concise, and organized reports in a way to gain the attention of the reader.

Technical knowledge and expertise relate to the individual's technical abilities to participate in the project. Further technical training will depend on the assignment's demands as well the participants' self-improvement motivation.

As the project progresses, the training staff will be setting up workshops and seminars for sales personnel. The training staff will also be involved with manufacturing and quality training.

QUALITY PLAN

Today, businesses place quality departments on the same organizational plane as other major departments. They are part of the team that influences the direction of the company. Quality takes on two aspects in everyday operations:

- **Quality assurance** establishes performance standards, measuring and evaluating performance to these standards and taking action when performance deviates from the standards.
- **Quality control** is the action taken to meet standards through monitoring, inspecting, testing, and gathering the performance information.

Quality becomes significant in a product development program because it affects four areas:

- customer satisfaction—satisfying a customer whose needs are consistent with the product that conforms strictly to its specifications
- design quality—setting targets and measuring the potential performance of the product
- process quality—developing defect prevention methods using statistical process control; developing product specifications
- manufacturing quality—developing quality control methods and procedures; continuous improvement through statistical process control and statistical quality control

CONCLUSION

When a company uses a phased approach with milestones to plan a product development program, it is likely to succeed because everyone will understand better what they are to accomplish. By dividing the total program into a series of steps, the project team can concentrate on selected activities within a phase, avoiding divided attention on a number of activities in the total program. This approach allows for less uncertainty.

Phasing projects of all classifications is part of the project management philosophy. Participants trained in project management techniques for organizing the product development project have a better understanding of the total program. Benefits include the following:

- *A proven procedure for communicating what is to be done in the project.* Correct communication with others involved in the project is important for its success.
- *A disciplined basis for planning a project.* Following the same course of action will avoid conflict, additional costs, and harmful delays.
- *A means to designate persons to assume responsibility to complete specific jobs in the project.* It identifies, emphasizes, and portrays individual

roles assigned to this project. The planning diagram is especially effective in showing their association and dependency upon other individuals.

- *Preparing the planning diagram provides a clear picture of the project.* Newly appointed personnel, after reviewing the diagram, will have a good understanding of what is to be done.
- *The planning diagram defines customer/supplier relationships.* It can specifically display the responses to the wants and needs of the customer.
- *An excellent way to evaluate alternate strategies and objectives.* Simulation studies using the plan and computerized schedules can be done quickly and accurately.

Exercises

1. Refer to Figure 11.5, the phase II WBS. Product design starts on January 23, 1997, the day after marketing submits the feasibility report to Dale, Inc. Four weeks later, on February 20, 1997, the Dale, Inc. board of directors approves the product development project. The product is to be introduced no later than March 16, 1999, about a 30-month program.

    The project team then tentatively established the following durations for each phase:

Phase I: Program Concepts	19 weeks (completed)	
Phase II: Product Design	24	"
Phase III: Product Development	19	"
Phase IV: Product Test	44	"
Phase V: Production	24	"

    a. Assuming that you are knowledgeable about how the activities in phase II relate to each other, prepare the project planning diagram for phase II.
    b. Assume that you are the project leader. Your responsibility is to determine the durations for each activity in phase II. (Before finalizing these durations, make a quick check to be sure the duration of the longest path is within the established schedule of phase II. If not, make adjustments until the objectives are satisfied.)
    c. What is the *Complete phase II* milestone date?
2. Refer to Figure 11.6, phase III WBS, and (1) prepare the project planning diagram and (2) determine durations for each activity for phase III. (Follow Exercise 1 directions.)
    a. Using manual calculations, determine the date for the *Complete phase III* milestone.
3. Complete planning diagrams and determine durations for activities for phases IV (Figure 11.7) and V (Figure 11.8), respectively.
    a. Based on the durations set up for the activities in phase IV and phase V, what are the durations for these phases?

Product Development Project

Phase II: Product Design

Marketing	Product Design/ Development	Engineering/ Manufacturing	Quality	Administrative/ Training
1. Prepare project approval documents	1. Define the product	1. Determine manufacturing processes	1. Assess reliability of the process	1. Prepare project plan and schedule

2. Secure project approval	2. Prepare product designs	2. Analyze process capabilities		2. Determine resources requirements
				  a. People
				  b. Materials
				  c. Equipment
				  d. Finances

	3. Prepare product specifications	3. Prepare facilities action plan		
	4. Authorize long-lead tooling and equipment	4. Determine facilities costs		

FIGURE 11.5 Phase II: Product design WBS for the product development project

271

Product Development Project
Phase III: Product Development

Marketing	Product Design/ Development	Engineering/ Manufacturing	Quality	Administrative/ Training
1. Decide sales strategy	1. Finalize design	1. Design tooling	1. Prepare quality plan	1. Prepare and implement project control procedures
2. Decide distribution strategy	2. Order long-lead tooling and equipment	2. Design manufac-turing facilities	2. Define process specifications	2. Select and qualify manufacturing personnel
3. Decide advertising strategy	3. Select prototype sources		3. Certify tooling and gages	

FIGURE 11.6 Phase III: Product development WBS for the product development project

Product Development Project

Phase IV: Product Test

Marketing	Product Design/ Development	Engineering/ Manufacturing	Quality	Administrative/ Training
1. Prepare sales manuals	1. Build prototype	1. Order tooling	1. Certify tooling, gages	1. Train manufacturing personnel
2. Prepare promotional literature	2. Test product	2. Build and install facilities	2. Measure prototype process performance	2. Train quality personnel
3. Prepare releases to media			3. Measure capability	3. Train maintenance personnel
4. Design advertising				

FIGURE 11.7 Phase IV: Product test WBS for the product development project

Product Development Project

Phase V: Production

Marketing	Product Design/ Development	Engineering/ Manufacturing	Quality	Administrative/ Training
1. Distribute sales manuals		1. Install tooling	1. Certify the production process	1. Train marketing personnel
2. Distribute promotional literature		2. Launch product	2. Qualify the product	2. Train customer service personnel
3. Release promotion to media		3. Start shipments	3. Finalize quality control methods	
4. Release advertising			4. Finalize quality control procedures	
			5. Finalize gage certification	
			6. Finalize tooling certification	

FIGURE 11.8   Phase V: Production WBS for the product development project

4. Dale, Inc.'s board of directors has set the following dates to review each phase of the product development project (which are to coincide with their completion dates):

Phase I (completed) January 22, 1997
Phase II July 9, 1997
Phase III November 19,1997
Phase IV September 23, 1998
Phase V March 10, 1999

a. Compare the "Complete" milestones of each phase that have been set in Exercises 1, 2, and 3 using the board's directive. Use this tabulation and fill in the blank spaces:

Comparison of Milestones

Milestone	Board	Project Team
Complete phase I	1/22/97	1/22/97
Complete phase II	7/9/97	_____
Complete phase III	_____	_____
Complete phase IV	_____	_____
Complete phase V	_____	_____

b. List the discrepancies between the milestones set by the board and those set by the project team. Which activity durations would you consider adjusting to match the milestones?

5. After completing the milestones for each phase and developing compatible timing data for each activity, produce a complete computerized schedule, sorted by phases, for the product development project.

Additional Instructions for IN CONTROL! Software

GENERATING REPORTS

The reporting options are all menu driven.

1. Begin at the Main menu.

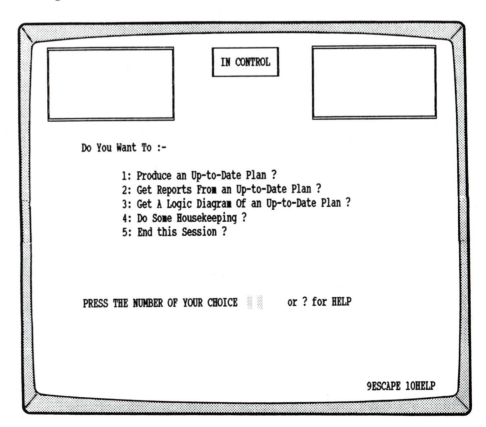

```
                          ┌──────────────┐
  ┌──────────────┐        │  IN CONTROL  │        ┌──────────────┐
  │              │        └──────────────┘        │              │
  │              │                                │              │
  │              │                                │              │
  └──────────────┘                                └──────────────┘


     Do You Want To :-

             1: Produce an Up-to-Date Plan ?
             2: Get Reports From an Up-to-Date Plan ?
             3: Get A Logic Diagram Of an Up-to-Date Plan ?
             4: Do Some Housekeeping ?
             5: End this Session ?




     PRESS THE NUMBER OF YOUR CHOICE          or ? for HELP



                                               9ESCAPE 10HELP
```

2. Press **2:** Get Reports from an Up-To-Date Plan?

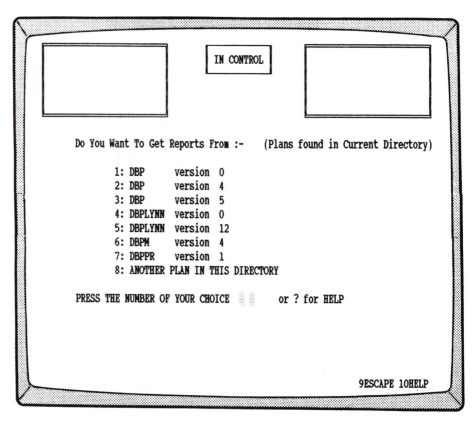

3. Press the number option of the plan you want to report on.
   (Select **8:** Another Plan in This Directory to see if your plan is on the listing on the next screen display.)
4. After you select a plan to report on, the reporting menu appears, presenting a list of options for project reporting.

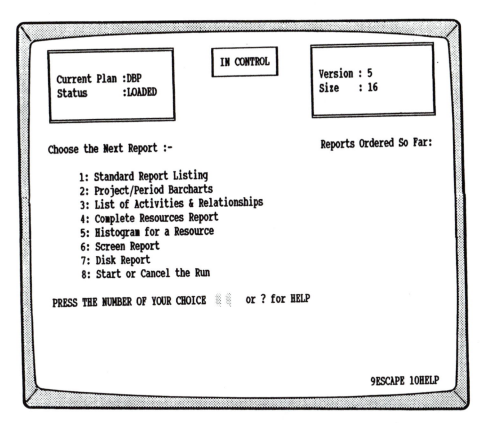

(This display identifies your plan, showing your reporting options and the order of reports you have ordered so far—note "Reports Ordered So Far" on right-hand block.)

Standard Report Listings

Standard report listings are printed versions of the tabulated schedule.
1. Press **1:** Standard Report Listing to generate a printed version of the tabulated schedule.
2. Press **1:** The Earliest Activities First? to have your reports generated from earliest starting date to latest starting date.
3. Press [ENTER] to the prompt Selection Key.
4. Press [ENTER] to the prompt Starting Location.
5. At the report menu, press **8**.
6. Press [ENTER] and the report shown in Figure A.1 will be printed:

```
PROJECT MANAGEMENT                              DESIGN/BUILD PROJECT                              BOCA RATON, FLORIDA
    REPORT TYPE :STANDARD LISTING                                         PRINTING SEQUENCE :Earliest Activities First
                                                                          SELECTION CRITERIA :ALL
    PLAN I.D.   :DBP      VERSION  1                                      TIME NOW DATE      : 2/OCT/95
```

ACTIVITY DESCRIPTION	EARLIEST START	EARLIEST FINISH	LATEST START	LATEST FINISH	DURATION	FLOAT
1- 4 DESIGN MECHANICAL EQUIPMENT	2/OCT/95	12/NOV/95	2/OCT/95	12/NOV/95	42	0 *
1- 3 DESIGN BUILDING	2/OCT/95	31/DEC/95	30/OCT/95	28/JAN/96	91	28
1- 2 DESIGN STRUCTURAL STEEL	2/OCT/95	15/OCT/95	6/NOV/95	19/NOV/95	14	35
2- 3 FABRICATE & ERECT STRUCTURAL STEEL	16/OCT/95	24/DEC/95	20/NOV/95	28/JAN/96	70	35
4- 5 DESIGN ELECTRICAL EQUIPMENT	13/NOV/95	10/DEC/95	13/NOV/95	10/DEC/95	28	0 *
4- 6 FABRICATE MECHANICAL EQUIPMENT	13/NOV/95	14/JAN/96	27/NOV/95	28/JAN/96	63	14
4- 8 DESIGN INTERIOR ITEMS	13/NOV/95	26/NOV/95	1/JAN/96	14/JAN/96	14	49
8- 9 PROCURE INTERIOR ITEMS	27/NOV/95	24/DEC/95	15/JAN/96	11/FEB/96	28	49
5- 6 PROCURE ELECTRICAL EQUIPMENT	11/DEC/95	28/JAN/96	11/DEC/95	28/JAN/96	49	0 *
9- 10 INSTALL INTERIOR ITEMS	25/DEC/95	3/MAR/96	12/FEB/96	21/APR/96	70	49
3- 7 CONSTRUCT PHASE I BUILDING	1/JAN/96	11/FEB/96	29/JAN/96	10/MAR/96	42	28
6- 7 INSTALL MECH & ELECT EQUIP	29/JAN/96	10/MAR/96	29/JAN/96	10/MAR/96	42	0 *
7- 11 CONSTRUCT PHASE II BUILDING	11/MAR/96	21/APR/96	11/MAR/96	21/APR/96	42	0 *

FIGURE A.1

Alphabetical Order

The alphabetical order shows the activities listed alphabetically by activity names. When used for the work breakdown structure (WBS), input alphabetical letters before each activity according to its position in the WBS.

1. Press **2:** The Output in Alphabetical Sequence?
2. Press **11** at *Starting Location.*
3. Press **1** at *Number of Characters.* (Use the appropriate figure for the number of alphabetical letters you are using.)

Critical Activities

The critical activities report places the activities in order according to the number of days of float associated with each activity.

1. Press **3:** The Most Critical Activities First?
2. Follow steps 3–6 of Standard Report Listing and the report shown in Figure A.2 will be printed:

```
PROJECT MANAGEMENT                          DESIGN/BUILD PROJECT                              BOCA RATON, FLORIDA
     REPORT TYPE :STANDARD LISTING                                    PRINTING SEQUENCE  :Most Critical Activities First
                                                                     SELECTION CRITERIA :ALL
     PLAN I.D.  :DBP     VERSION 5                                    TIME NOW DATE      : 2/OCT/95
===================================================================================================================
     ACTIVITY DESCRIPTION               EARLIEST    EARLIEST    LATEST      LATEST     DURATION FLOAT
                                        START       FINISH      START       FINISH
===================================================================================================================
   1-  4 DESIGN MECHANICAL EQUIPMENT      2/OCT/95   12/NOV/95    2/OCT/95   12/NOV/95     42      0 *
   4-  5 DESIGN ELECTRICAL EQUIPMENT     13/NOV/95   10/DEC/95   13/NOV/95   10/DEC/95     28      0 *
   5-  6 PROCURE ELECTRICAL EQUIPMENT    11/DEC/95   28/JAN/96   11/DEC/95   28/JAN/96     49      0 *
   6-  7 INSTALL MECH & ELECT EQUIP      29/JAN/96   10/MAR/96   29/JAN/96   10/MAR/96     42      0 *

   7- 11 CONSTRUCT PHASE II BUILDING     11/MAR/96   21/APR/96   11/MAR/96   21/APR/96     42      0 *
   4-  6 FABRICATE MECHANICAL EQUIPMENT  13/NOV/95   14/JAN/96   27/NOV/95   28/JAN/96     63     14
   1-  3 DESIGN BUILDING                  2/OCT/95   31/DEC/95   30/OCT/95   28/JAN/96     91     28
   3-  7 CONSTRUCT PHASE I BUILDING       1/JAN/96   11/FEB/96   29/JAN/96   10/MAR/96     42     28

   1-  2 DESIGN STRUCTURAL STEEL          2/OCT/95   15/OCT/95    6/NOV/95   19/NOV/95     14     35
   2-  3 FABRICATE & ERECT STRUCTURAL STEEL 16/OCT/95 24/DEC/95  20/NOV/95   28/JAN/96     70     35
   4-  8 DESIGN INTERIOR ITEMS           13/NOV/95   26/NOV/95    1/JAN/96   14/JAN/96     14     49
   8-  9 PROCURE INTERIOR ITEMS          27/NOV/95   24/DEC/95   15/JAN/96   11/FEB/96     28     49

   9- 10 INSTALL INTERIOR ITEMS          25/DEC/95    3/MAR/96   12/FEB/96   21/APR/96     70     49
===================================================================================================================
```

FIGURE A.2

Project/Period Bar Charts

Bar charts are a graphic display of the information generated in the standard report. The project bar chart covers the whole project, while the period bar chart covers only a portion of it.

1. Press **2:** Project/Period Barcharts.
2. Press **1:** The Earliest Activities First.
3. Press [ENTER][ENTER] to reach the Barchart report screen.
4. For the option, period bar chart: Type **2** at the prompt *Type the Day Number That You Wish This Report to Start From.*
5. Press [ENTER].
6. A period bar chart report is a daily printout that may extend over several pages, depending on the length of the project.

(If you respond with a specific working day number, you will obtain a period bar chart starting on that specified day and continuing until it fills one page width on your printer.)

To get more days per page, you may follow your working day entry with a number in parentheses that will indicate the number of days to represent per print position (e.g., the response **5(7)** would generate a period bar chart beginning at work day 5, and would include 7 days per print position).

7. For the option, project bar chart: Type **0(0)**.
8. Press [ENTER].
9. A project bar chart is a printout of a total project compressed on one page. This project bar chart report (see Figure A.3) shows the earliest start activities first:

```
PROJECT MANAGEMENT                              DESIGN/BUILD PROJECT                              BOCA RATON, FLORIDA
        REPORT TYPE :COMPRESSED PERIOD BARCHART                              PRINTING SEQUENCE  :Earliest Activities First
                                                                            SELECTION CRITERIA :ALL
        PLAN I.D.   :DBP      VERSION 0                                      TIME NOW DATE       : 2/OCT/95
===========================================1995========================1996=============================================
   PERIOD COMMENCING DATE              !2    !27    !21   !16    !10   !4    !29    !25    !19   !14    !8    !3    !
   MONTH                               !OCT  !      !NOV  !DEC   !JAN  !FEB  !      !MAR    !APR  !MAY   !JUN  !JUL  !
   PERIOD COMMENCING TIME UNIT         !2    !27    !52   !77    !102  !127  !152   !177    !202  !227   !252  !277  !
======================================================================================================================
   1-   4 DESIGN MECHANICAL EQUIPMENT  !CCCCCC!CCCCC !     !      !     !     !      !      !     !      !     !     !
   1-   3 DESIGN BUILDING              !======!======!======!====..!......!     !      !      !     !      !     !     !
   1-   2 DESIGN STRUCTURAL STEEL      !====..!......!.     !      !     !     !      !      !     !      !     !     !
   2-   3 FABRICATE & ERECT STRUCTURAL S !  ===!======!======!===....!......!     !      !      !     !      !     !     !
   ------------------------------------------------------------------------------------------------------------------
   4-   5 DESIGN ELECTRICAL EQUIPMENT  !     !     CC!CCCCCC!      !     !     !      !      !     !      !     !     !
   4-   6 FABRICATE MECHANICAL EQUIPMENT !    !     ==!======!======!==... !     !      !      !     !      !     !     !
   4-   8 DESIGN INTERIOR ITEMS        !     !     ==!==....!......!..     !     !      !      !     !      !     !     !
   8-   9 PROCURE INTERIOR ITEMS       !     !     !    ====!===....!......!...     !      !      !     !      !     !     !
   ------------------------------------------------------------------------------------------------------------------
   5-   6 PROCURE ELECTRICAL EQUIPMENT !     !     !  C!CCCCCCCC!CCCCC !     !      !      !     !      !     !     !
   9-  10 INSTALL INTERIOR ITEMS       !     !     !     !    ====!======!======!==....!......!.     !      !     !     !
   3-   7 CONSTRUCT PHASE I BUILDING   !     !     !     !   ===!======!===...!.....  !      !     !      !     !     !
   6-   7 INSTALL MECH & ELECT EQUIP   !     !     !     !      !    CC!CCCCCC!CCCC  !      !     !      !     !     !
   ------------------------------------------------------------------------------------------------------------------
   7-  11 CONSTRUCT PHASE II BUILDING  !     !     !     !      !     !     !   CCC!CCCCCCC!C   !      !     !     !
======================================================================================================================
Barchart Key:-  CCC :Critical Activities   === :Non Critical Activities   NNN :Activity with neg float   ... :Float
```

FIGURE A.3

10. At Item 2, if you had pressed **3:** The Most Critical Activities First, then
followed the same sequence as above, you would get the report shown in
Figure A.4:

```
PROJECT MANAGEMENT                              DESIGN/BUILD PROJECT                               BOCA RATON, FLORIDA
       REPORT TYPE :COMPRESSED PERIOD BARCHART                                  PRINTING SEQUENCE :Most Critical Activities First
                                                                               SELECTION CRITERIA :ALL
       PLAN I.D.   :DBP     VERSION 5                                           TIME NOW DATE      : 2/OCT/95
==============================================1995=====================1996============================================
   PERIOD COMMENCING DATE          !2     !27    !21    !16    !10    !4     !29    !25    !19    !14    !8     !3     !
   MONTH                           !OCT   !      !NOV   !DEC   !JAN   !FEB   !   .  !MAR   !APR   !MAY   !JUN   !JUL   !
   PERIOD COMMENCING TIME UNIT     !2     !27    !52    !77    !102   !127   !152   !177   !202   !227   !252   !277   !
====================================================================================================================
 1-   4 DESIGN MECHANICAL EQUIPMENT  !CCCCCC!CCCCC !      !      !      !      !      !      !      !      !      !      !
 4-   5 DESIGN ELECTRICAL EQUIPMENT  !      ! CC!CCCCCC!     !      !      !      !      !      !      !      !      !
 5-   6 PROCURE ELECTRICAL EQUIPMENT !      !      !      C!CCCCCCC!CCCCC !      !      !      !      !      !      !
 6-   7 INSTALL MECH & ELECT EQUIP   !      !      !      !      !      CC!CCCCCC!CCCC  !      !      !      !      !

 7-  11 CONSTRUCT PHASE II BUILDING  !      !      !      !      !      !      !      CCC!CCCCCCC!C    !      !      !
 4-   6 FABRICATE MECHANICAL EQUIPMENT !    !      ==!======!======!==... !      !      !      !      !      !      !
 1-   3 DESIGN BUILDING              !======!======!======!======..!.... !      !      !      !      !      !      !
 3-   7 CONSTRUCT PHASE I BUILDING   !      !      !      !      ===!======!===...!..... !      !      !      !      !

 1-   2 DESIGN STRUCTURAL STEEL      !====..!.......!.    !      !      !      !      !      !      !      !      !
 2-   3 FABRICATE & ERECT STRUCTURAL S !  ===!======!======!===....!.... !      !      !      !      !      !      !
 4-   8 DESIGN INTERIOR ITEMS        !      !      ==!==....!.......!.    !      !      !      !      !      !      !
 8-   9 PROCURE INTERIOR ITEMS       !      !      ! ====!===....!.......!.... !      !      !      !      !      !

 9-  10 INSTALL INTERIOR ITEMS       !      !      !      !  ====!======!======!==....!.......!.   !      !      !
====================================================================================================================
Barchart Key:-  CCC :Critical Activities  === :Non Critical Activities   NNN :Activity with neg float  ... :Float
```

FIGURE A.4

List of Activities and Relationships

This report prints out descriptions of and resources for the activities—useful for
"debugging" your project plan.

1. Press **3:** List of Activities & Relationships.
2. This generates a report entitled Network Listing under your Reports Ordered
So Far.

If you want this report, select it first, since the project data have not yet been
affected by analyses that may accompany other reporting options.

MAKING CHANGES

Since plans are dynamic, changes are usually necessary as time goes by. For data
already input, there is a varied menu of change options.

General Directions

1. Bring up Main menu.
2. Press **1:** Produce an Up-to-Date Plan.
3. Press option number of plan to be changed.
4. The first 19 lines of that plan will be displayed. Always type the line number
of the activity that you are changing.

Changing Activity Durations

Changes made most often will be revisions to the activity durations.

1. At *Command*, type the line number of the activity whose duration is to be changed.
2. The cursor will move to the Duration column. Type the new duration and press [ENTER].
3. Press [ENTER] again and the cursor returns to the Command line without disturbing the Description column.

Changing Descriptions

1. At *Command*, type the line number of the activity whose description is to be changed.
2. Press [ENTER] at *Duration* column.
3. Type necessary changes to existing description.
4. Move cursor to end of description. Press [ENTER] to return to Command.

Adding Activities

Adding activities to your project plan is a common occurrence once you have completed the plan.

1. Type **I** on the Command line (or press **F3 Insert** shown at the bottom of the input screen).
2. Start typing node numbers of the activity you wish to add.
3. Continue in the same manner as you had for adding activities in the original plan.
4. Press [ESC] after you have entered the additional activities.
5. The added activities will be incorporated in the original plan.

Changing the Identification

You can change information about your project—its name, the type of project, calendar and abbreviations files, time units, and report titles—without reentering all of the data. The only exception is that you cannot change from an arrow to a precedence plan (or vice versa), since data are entered differently for the two plans.

1. At the *Command* line, type **H** (or press **F5**).
2. Move the cursor by pressing [ENTER] to get to the item to be changed.
3. Type over existing words or use the space bar to eliminate characters at the end.

Suggestion: If you have been working on many versions of a plan, and the version number is quite high, simply type over the name of the plan with the same name and it will reset the version number to zero.

Delete Activity

Revisions may also require deleting an activity. As with other changes to the input, use the line number to refer to the activity.

1. Type **D** followed by the line number of the activity to be deleted.
2. Press [ENTER], then press [ENTER] again to confirm the deletion.
3. For example, type **D2** to delete the activity on line 2. Press [ENTER], then press [ENTER] again.
4. The screen will clear and redisplay 19 activities with the deleted line gone.
5. Press [ESC] before the second [ENTER] to cancel the delete request.

ADDITIONAL FEATURES IN THE HOUSEKEEPING MENU

Several items that may be of benefit in the use of this software are:

1. Removing unwanted plans.
2. Copying plans from one drive to another.
3. Selected printer details.
4. Selected system details.

Removing Unwanted Plans

Since this software changes project versions consecutively, with every change it may be prudent to rid the files of plans and/or versions of plans periodically.

1. From main menu press **4:** Do Some Housekeeping?

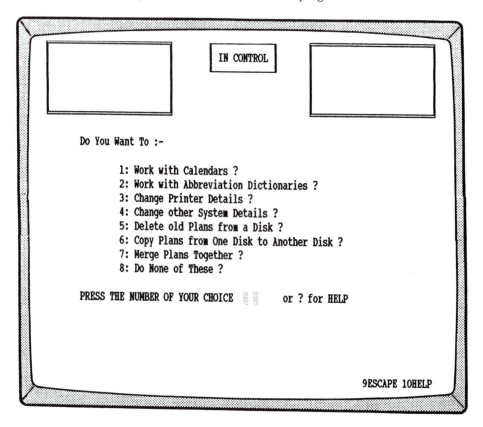

```
                          ┌──────────────┐
                          │  IN CONTROL  │
                          └──────────────┘


   Do You Want To :-

              1: Work with Calendars ?
              2: Work with Abbreviation Dictionaries ?
              3: Change Printer Details ?
              4: Change other System Details ?
              5: Delete old Plans from a Disk ?
              6: Copy Plans from One Disk to Another Disk ?
              7: Merge Plans Together ?
              8: Do None of These ?

   PRESS THE NUMBER OF YOUR CHOICE           or ? for HELP



                                              9ESCAPE 10HELP
```

2. Press **5:** Delete Old Plans from a Disk?
3. Press line number of plan to be deleted.
4. Press **Y** to confirm that plan is to be deleted.
5. Press **Y** if you wish to continue deleting; press **N** to stop deleting.

Copying Plans

Plans should be stored on a diskette to avoid possible erasing while using the program. Copying from one drive to the hard drive and vice versa is relatively easy.

1. From main menu press **4:** Do Some Housekeeping?
2. Press **6:** Copy Plans from One Disk to Another Disk?
3. Respond to The Drive to be Copied From? and then The Drive to be Copied To?

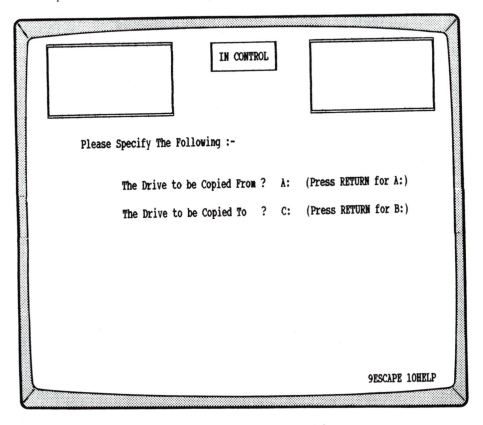

```
                            ┌──────────────┐
                            │  IN CONTROL  │
                            └──────────────┘



        Please Specify The Following :-


            The Drive to be Copied From ?   A:   (Press RETURN for A:)

            The Drive to be Copied To   ?   C:   (Press RETURN for B:)




                                                      9ESCAPE 10HELP
```

4. Select line number corresponding to plan you wish to copy.

Configuring System: Printer Details

This is a precautionary check as the default settings may already be compatible for the printer you will be using.

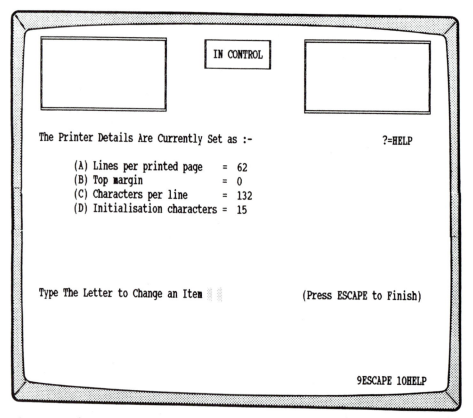

1. Press **A** for the paper length. Usually 66 lines per page is appropriate for portrait size (8.5 by 11); 45 lines per page for landscape (11 by 8.5).
2. Press [ENTER].
3. Press **B** for desired top margin. (0 provides an acceptable top margin.)
4. Press [ENTER].
5. Press **C** for the number of characters per line. This must be a minimum of 132 and a maximum of 252 characters. (Check your printer manual for further details.)
6. Press [ENTER].
7. Press **D** to control special features. (For example, your printer may not respond to compressed printing for this program even though there is a compressed font on the printer control. In this event, refer to the printer manual for the proper sequence of decimal numbers. Separate the sequence with diagonals.)
8. Press [ENTER].
9. When complete, press [ESC] to return to main menu.

Configuring System: System Details

This feature permits placing data directly on the drive you desire and allows for customizing reports.

   1. From the main menu, press **4:** Do Some Housekeeping?
   2. Press **4:** Change Other System Details?

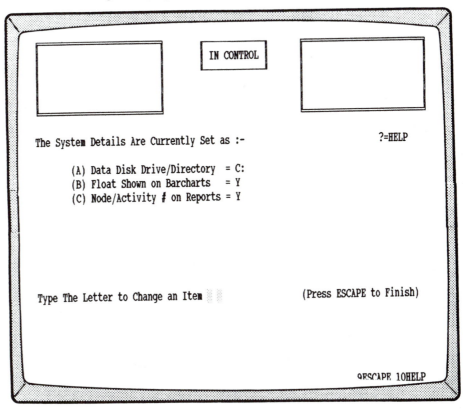

```
                              ┌──────────────┐
                              │  IN CONTROL  │
                              └──────────────┘

     The System Details Are Currently Set as :-                    ?=HELP

          (A) Data Disk Drive/Directory  = C:
          (B) Float Shown on Barcharts   = Y
          (C) Node/Activity # on Reports = Y




     Type The Letter to Change an Item              (Press ESCAPE to Finish)



                                                         9ESCAPE 10HELP
```

   3. Press **A** for drive where you wish your data to reside.
   4. Press [ENTER].
   5. Press **B** for the option of showing the float on bar chart reports. (If a bar chart is given to vendors, suppliers, or contractors you may not wish them to know about optional dates to deliver their services. Type **N** to show no float.)
   6. Press [ENTER].
   7. Press **C** for the option of including node and/or activity numbers on reports. (For example, type **N** to remove the node/activity number for reports to management—they have no need for the node and/or activity numbers, which then allows more space for the activity descriptions.)
   8. Press [ENTER].
   9. When complete, press [ESC] to return to main menu.

WORKSHEETS

Before beginning to enter data for the sample project, it is useful to use worksheets. These sheets will contain the data from the project.

Abbreviations Worksheet

Abbreviations codes for the various resources are needed for this program, so it is helpful to prepare an abbreviations tabulation. Normally all abbreviations are two-character codes. Exceptions are: (1) For an accumulated cost curve, the first character is a dollar sign ($), and (2) if the first character is a pound sign (#), then the resource is treated as a point resource and the costs will not be spread out over the duration of that activity. See Figure A.5 for the abbreviation worksheet for the Christopher Design/Build Project.

FIGURE A.5

Abbreviations Worksheet

Project Name:
Christopher Design/Build Project

Abbreviations File Name (<8 Characters): DBPA		
Resource Descriptions (<38 Characters)	**Conventional or Point(#)**	**Abbr***
Skilled personnel/Labor	Conventional	SP
Designers	Conventional	DE
Architects	Conventional	AR
Engineers	Conventional	EN
Structural		SL
Building		BU
Mechanical		ML
Electrical		EL
Architectural		AL
Project activity costs	Conventional	TC
Accumulative costs		$C
*2 Characters required		

FIGURE A.6

Calendar Worksheet

Project Name:
Christopher Design/Build Project

Calendar file name (<8 Characters): DBPC
First day (Mth/day/yr): Oct/2/95
Work days: MON[X]   TUES[X]   WED[X]   THU[X]   FRI[X]   SAT[X]   SUN[X]
Special non-work days:
Special work days:

Calendar Worksheet

Since the design/build project will be a "crash project," the calendar information will show Saturdays and Sundays as work days. In fact, there will be no nonwork days, including holidays (see Figure A.6).

Identification "Scorecard"

Before inputting time and resource data, the program requires some fundamental information for identifying input data files and printed reports. The identification scorecard includes the file name, report titles, calendar and abbreviations file names, plan format (arrow or precedence), and time units per day. Time units per day allow for recognition of multiple shifts in a day (see Figure A.7).

FIGURE A.7

Identification Worksheet

Project Name:
Christopher Design/Build Project

File name for project (<8 Characters):	DBPC
Title: top left of report (<30 Characters):	Project management
Title: top center of report (<30 Characters):	Design/build project
Title: top right of report (<30 Characters):	Boca Raton, Florida
Calendar file name (<8 Characters):	DBPC
Abbreviations file name (<8 Characters):	DBPA
Plan format type (A=Arrow, P=Precedence)	A
Time units per day (<3 Digits)	I

Activities/Resources Worksheet

The activities worksheet is set up to ease the inputting process. It will include data for use in both the timing input and the resources input screens. Resource levels must be shown on a daily basis, except for the point resource, #M. Milestone events are also included in this worksheet. Note that the numbered nodes for the activities and milestones need not be consecutive. The activities worksheet for the sample project is shown in Figure A.8.

Activities/Resources Worksheet

Project Name:
Christopher Design/Build Project

Start Node	End Node	Duration (Days)	Description	Resources (Daily Requirements)			
				EN	AR	DE	SP
Start	1	—	————	—	—	—	—
1	2	14	Design structural steel	2	—	6	—
1	3	91	Design building	6	4	12	—
1	4	42	Design mechanical equipment	4	—	4	—
2	3	70	Fabricate and erect structural steel	1	—	2	10
3	7	42	Construct phase I building	3	—	—	20
4	5	28	Design electrical equipment	4	—	8	—
4	6	63	Fabricate mechanical equipment	2	—	4	—
4	8	14	Design interior items	—	4	8	—
5	6	49	Procure electrical equipment	2	—	1	—
6	7	42	Install mechanical and electrical equipment	4	—	—	16
7	11	42	Construct phase II building	5	—	—	20
8	9	28	Procure interior items	—	2	2	—
9	10	70	Install interior items	—	2	—	12
10	11	—	Dummy	—	—	—	—
11	F	—	————				

1	Milestone	Start project	
3	Milestone	Complete structural steel	EN—Engineers
6	Milestone	Complete electrical equipment–	AR—Architects
		design and procure	DE—Designers
7	Milestone	Complete mechanical and electrical;	SP—Skilled Personnel/Labor
		Complete phase I	
10	Milestone	Complete architectural	
11	Milestone	Complete project	

FIGURE A.8

Greenfield Facility Description

The general outline of the property, configuration of the building, and allocation of exterior and interior space for required facilities and services are indicated on the Site Plan. A description of the requirements necessary for this manufacturing facility follows.

A: SITE

The area required is approximately 35 acres, which should be relatively flat, have good drainage, have good access to major highways, and be near a city capable of providing an annual domestic water supply of 149 million gallons and a sanitary sewage treatment system for handling the plant discharges. Additional utilities required are 358 million cubic feet of natural gas and an electric supply of 74 million KVH. Demand requirements are 16,800 KW of electricity, 405 GPM of water, and 116,300 CFH of natural gas.

B: BUILDING

The building shall be a single floor, with partial mezzanine levels, a structural steel framed, metal-sided structure having a roof area of approximately 355,400 square feet with potential to expand to 497,500 square feet. Generally, all bays shall be 40 feet, 0 inches by 50 feet, 0 inches, with an eave height of 26 feet (minimum). Building design may utilize prefabricated building technology with minimum roof loads to meet process requirements.

C: FACILITIES AND SERVICES

To qualify as a minimally acceptable facility, the main structure shall have sufficient space to permit the allocation of areas as follows:

Area	Square Footage
Manufacturing	243,800
Second floor process area	41,900
Stock storage	35,800
Packing and warehouse	62,400
Maintenance and tool manufacturing	5,400
Quality testing	8,600
Total	397,900

A separate structure consisting of 38,750 square feet is needed for administrative and employee facilities. Utilities will be located in a separate area including a main electrical substation, cooling water treatment, compressed air generating equipment, and other utilities.

Consideration is also to be given to other facilities and services required for plant start-up. These include:

- natural gas supply including gas meter station from local utility
- domestic water supply from local municipality
- fire protection system, including perimeter fire loop, fire water storage tank, pumps, fire hose stations, and sprinklers
- sanitary sewer system directed to the local municipality for treatment
- compressed air system
- hot water supply for domestic use
- heating (gas fired), ventilating, and air conditioning appropriately designed for specific areas
- electrical, including primary power service, secondary power distribution, lighting, telephone, fire alarm, and security
- site features, including parking lots, roadways, fencing, and outside lighting

Any of the above items not included in an existing property must have the capability of being added on or incorporated in accordance with reasonable and acceptable engineering practices. Specific details regarding the auxiliary facilities and services will be included in the user requirements as soon as the information becomes available. An investigation will be made of possible environmental implications.

"Greenfield" Facility
Site Plan

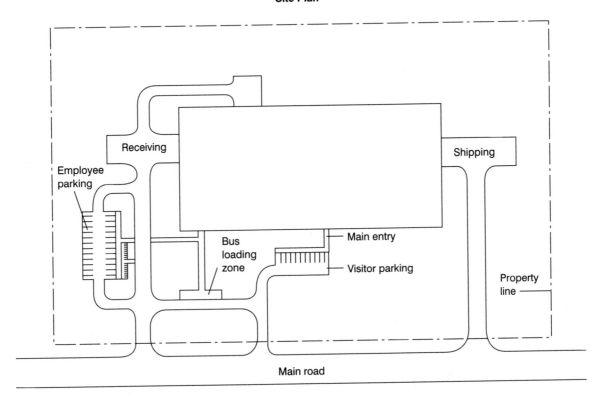

Receiving

Employee
parking

Shipping

Bus
loading
zone

Main entry

Visitor parking

Property
line

Main road

New Plant Site Comparison

	Site A	Site B	Site C
I. Land Characteristic			
• Land	One owner (private)	One owner (private)	One owner (private)
• Acres available	520 Acres	220 acres, lease for 58 years or 99 years	188 Acres
• Land cost or rental cost	$6200/acre (estimated)	$240/acre/year (estimated)	$8,000–$10,000 (estimated)
• Land configuration	Good	Poor—plant would be close to road and below road level	Good
• Miscellaneous charges	None	Higher taxes depending on future action by port authority	None
II. Transportation			
• Railroad	Frisco Railroad 2 miles south of site. The rail line can be extended; however, it will require excessive rock excavation.	• Serviced by two railroads—Frisco and Santa Fe service to site • Possibility of rail car movement by port authority on site, which could cause union problems	• MKT railroad service adjacent to site

Site Selection
Nonquantifiable Criteria
Summary

Criteria	Weight	Site A	Site B
Site Requirements	5		
Transportation	10		
Utilities	5		
Supply	20		
Labor	35		
Local Environment	35		
Subtotal	90		
Governmental (incentives)	10		
Total	100		

Area Evaluation Matrix
Nonquantifiable Criteria Table

Parameters	Weight	Site A Rating	Site A Score	Site B Rating	Site B Score
Site requirements					
• Topography (flood plain)	4				
• Site configuration	2				
• Surrounding sites	3				
• Environmental considerations	5				
• Climatological conditions	2				
• Seismic zone	4				
• Soils and foundation	4				
Total score: Weight × Rating					
Transportation					
• Proximity to major city	4				
• Proximity to markets	2				
• Warehousing at customer	5				
• Airport proximity	5				
• Rail facilities	0				
• Main highway proximity	5				
• Feeder roadways	4				
• Highway conditions (reliability)	3				
• Trailer staging availability	4				
Total score: Weight × Rating					
Total Score: Site Req., Trans.					

Concept-to-Market (CTM) Project Management Seminar Outline

DAY ONE

Attending Functions: Marketing
Product design/development
Engineering/manufacturing
Quality
Administrative/training

I. Introduction
   A. Seminar procedure, objectives, and scope
   B. Overview of project management training; roles of project manager and team members
   C. Why project management: applications and benefits
II. Planning a Project
   A. Planning procedures
      1. Review major objectives, milestones
      2. Review CTM functional departments and phases
      3. Designate assignments—team leaders, members
      4. Overview of planning technique
      5. Construct work breakdown structure (WBS) by each function for each phase
         a. Designate tasks (or activities)
         b. Identify responsibilities
         c. Determine resources
      6. Construct planning diagram
         a. Instruction on rules for constructing planning diagram
         b. Determine objectives and milestones for each phase

      c. Using WBS, determine relationships among each function for each phase

      d. Construct functional planning diagrams for each phase
         (1) Determine activity interrelationships
         (2) Assign timing to activities
         (3) Assign resources to activities

      e. Construct final planning diagram (assembled from the functional planning diagrams)

III. Wrap-Up—Day One

DAY TWO

I. Developing the Project Schedule
    A. Scheduling procedures
      1. Overview of scheduling technique
        a. Obtaining time estimates for activities
        b. Define *earliest, latest starting,* and *finishing dates; float times, critical activities,* and *"critical path"*
      2. Computerized scheduling
        a. Preparing input data from project planning diagram
          (1) Using worksheets
          (2) Input by sorts (or categories)
        b. Computer-produced timing reports
          (1) Screen reports
          (2) Schedule listings
          (3) Bar charts
          (4) Milestone reports
    B. Using computer for project analysis
      1. Sorting for customizing reports
      2. What-if options using the computer screen and edit feature
      3. Identify potential opportunities and/or potential problems using the float sort

II. Wrap-Up—Day Two

DAY THREE

I. Measuring Project Performance
    A. Procedures for project control
      1. Measuring performance: overview
      2. Monitor, assess, and resolve potential problems
      3. Communicating: meetings, memos, one-on-one
      4. Differences between *critical path management* and *report management*

        B.   Preparing status reports
            1. Using the float feature to control project with contractors, suppliers
            2. Using the float feature for management reports
  II. Planning Personnel Requirements
        A.   Procedure for personnel planning
            1. Personnel planning overview
            2. Identify the personnel required to complete each activity and mark on planning diagram
            3. Prepare computer input data (using worksheet)
            4. Select desired personnel loading reports
        B.   Evaluate personnel demand versus availability
            1. Produce what-if schedules to achieve leveling
            2. Analyze overdemand situations and offer solutions
  III. Critique
        A.   Concluding remarks
        B.   Seminar evaluation: Additional training required?

Index

About the Author

M. Pete Spinner is a project management consultant who has worked with Fortune 500 engineering, construction, and manufacturing organizations. He recently retired from Ford Motor Company after 35 years of service. While there, he managed engineering departments with responsibility for engineering, design, product planning and development, installation, construction, energy and environmental matters, maintenance planning, and project management. Mr. Spinner began applying project management techniques at Ford in 1962, an endeavor that evolved into his becoming the leading authority at Ford in applying project management techniques in facilities engineering and construction, product planning and development, and advanced engineering projects.

Mr. Spinner has developed and conducted seminars under the sponsorship of several universities, as well as training hundreds of Ford personnel, including personnel in engineering, manufacturing, computer systems, finance, training, and research and development. These seminars stressed improved methods of planning, scheduling, and organizing projects and programs.

Mr. Spinner is the author of *Elements of Project Management* and *Improving Project Management Skills and Techniques* and has presented numerous papers on the subject of project management to various professional organizations. He is a certified Project Management Professional and is registered as a Professional Engineer in Michigan. He is an active member of the Project Management Institute. He holds a Master of Science in Civil Engineering from Harvard University, a Bachelor of Science in Civil Engineering from the University of Pittsburgh, and completed the Army Specialized Training Program at Purdue University.